T0140309

Natural Computing Series

Founding Editor

Grzegorz Rozenberg

Series Editors

Thomas Bäck ⓘ, Natural Computing Group–LIACS, Leiden University, Leiden, The Netherlands

Lila Kari, School of Computer Science, University of Waterloo, Waterloo, ON, Canada

Susan Stepney, Department of Computer Science, University of York, York, UK

Scope

Natural Computing is one of the most exciting developments in computer science, and there is a growing consensus that it will become a major field in this century. This series includes monographs, textbooks, and state-of-the-art collections covering the whole spectrum of Natural Computing and ranging from theory to applications.

More information about this series at https://link.springer.com/bookseries/4190

Tome Eftimov · Peter Korošec

Deep Statistical Comparison for Meta-heuristic Stochastic Optimization Algorithms

 Springer

Tome Eftimov
Computer Systems Department
Jožef Stefan Institute
Ljubljana, Slovenia

Peter Korošec
Computer Systems Department
Jožef Stefan Institute
Ljubljana, Slovenia

ISSN 1619-7127
Natural Computing Series
ISBN 978-3-030-96919-6   ISBN 978-3-030-96917-2   (eBook)
https://doi.org/10.1007/978-3-030-96917-2

© The Editor(s) (if applicable) and The Author(s), under exclusive license to Springer Nature
Switzerland AG 2022
This work is subject to copyright. All rights are solely and exclusively licensed by the Publisher, whether
the whole or part of the material is concerned, specifically the rights of translation, reprinting, reuse
of illustrations, recitation, broadcasting, reproduction on microfilms or in any other physical way, and
transmission or information storage and retrieval, electronic adaptation, computer software, or by similar
or dissimilar methodology now known or hereafter developed.
The use of general descriptive names, registered names, trademarks, service marks, etc. in this publication
does not imply, even in the absence of a specific statement, that such names are exempt from the relevant
protective laws and regulations and therefore free for general use.
The publisher, the authors and the editors are safe to assume that the advice and information in this book
are believed to be true and accurate at the date of publication. Neither the publisher nor the authors or
the editors give a warranty, expressed or implied, with respect to the material contained herein or for any
errors or omissions that may have been made. The publisher remains neutral with regard to jurisdictional
claims in published maps and institutional affiliations.

This Springer imprint is published by the registered company Springer Nature Switzerland AG
The registered company address is: Gewerbestrasse 11, 6330 Cham, Switzerland

*To my mother, father, brother, both
grandmothers and* **both** *grandfathers for
giving me their unconditional support.
and
For the endless discussions about nothing
and everything ...*

—Tome

*To the family I grew up with: mother Jožica,
father Lado, and brother Borut
and
To the family growing up with me: my beloved
Tina, and our children, Tim, Nik, and Neli*

—Peter

Foreword

Optimization is finding the best solution from among all the feasible solutions. In mathematics and computer science, this challenging task is being tackled by many researchers around the world who are inventing various approaches and algorithms. In the past 20 years, the inventing of new optimization algorithms has become something of a competition, not only with oneself, but also with other researchers. In theory, competitive activities have many benefits and help researchers develop important skills, like perseverance and resilience, as well as protecting science from stagnation. However, to avoid unfair competition, clear rules need to be defined and agreed on when comparing the approaches and algorithms. Tome Eftimov and Peter Korošec are that rare breed of computer scientist that recognize the need to define the rules for benchmarking algorithms in a specific field of meta-heuristic stochastic optimisation. They provide a fascinating compendium of theoretical and practical guides and even a web-based tool for making a statistical comparison of both single-objective and multi-objective optimisation algorithms in a fair way. This book provides a path forward, not only in the field of meta-heuristic stochastic optimisation, but also in other fields of computer science where approaches and algorithms should be benchmarked in the correct way before announcing them as the best solutions. As a consequence, the researcher who wins the competition will succeed as a true scientist, which is the most beautiful feeling of all.

Ljubljana, Slovenia   Barbara Koroušić Seljak
November 2021

Preface

Everything started in 2015. I (Tome Eftimov) came back from a conference, and my Ph.D. supervisor (Barbara Koroušić Seljak) told me that it would be nice to make a presentation about how many runs we should make to obtain a representative data sample for a statistical analysis of meta-heuristics. For this, I needed to collaborate with my colleague Peter Korošec. The idea was to transfer the statistical knowledge applied in other domains to optimization, where the input was the experimental data from running single-objective meta-heuristics. After a few discussions about the statistical methods, we both realized that this could not be applied to our application scenario. Peter explained to me about the weaknesses when statistical comparisons are made using meta-heuristics data. After a few hours, I had implemented the Deep Statistical Comparison approach. The next day, after explaining this to Peter and discussing the results, he told me that we should publish them in a journal. This is how the Deep Statistical Comparison story began, with neither of us realizing at the time the benefits we would derive from making statistical comparisons more robust. Since then, there have been many long discussions, which have finally led to this book. And despite some of the discussions becoming heated, we are happy to consider ourselves as friends, not just colleagues.

Our initial study about how many optimization runs we would need to obtain a representative data sample for a statistical analysis of meta-heuristics remains unfinished. But our diversion into the Deep Statistical Comparison approach was worth it.

Book Content

This book explains what is required to make a more robust statistical comparison of the performance achieved by meta-heuristics. It does not explain how to develop a new single- or multi-objective meta-heuristic, or how to run it on a specific or a set of problem instances and collect the experimental data. It deals in which statistical analysis should be made once the experimental data is collected to obtain robust statistical outcomes.

Intended Audience

The book has been written with three audiences in mind:

- Students in the field of meta-heuristic stochastic optimization.
- Experienced researchers in the field of meta-heuristic stochastic optimization.
- Engineers who need to select an appropriate optimization algorithm for their industrial task based on its performance.

Expected Background Knowledge

Few assumptions are made relating to the background of the reader. However, they should know where to find meta-heuristic stochastic optimization algorithms, how to run them, the basics of how they work, and how the experimental data are collected. All this knowledge is not essential, but it will help in understanding the logic behind the statistical analysis presented in this book.

Book Outline

The book is organized for both new and experienced readers. For the new readers it provides the basics in optimization and statistical analysis, allowing them to become familiar with the material before the introduction of the Deep Statistical Comparison approach. Experienced researchers can move rapidly to the introduction of the new statistical approaches. To accommodate the reader, the content is organized into three parts:

- Part I: Introduction to optimization, benchmarking, and statistical analysis—Chaps. 2–4.
- Part II: Deep Statistical Comparison of meta-heuristic stochastic optimization algorithms—Chaps. 5–7.
- Part III: Implementation and application of the Deep Statistical Comparison—Chap. 8.

The aim of the first part is to provide the basics in optimization, benchmarking, and statistical analysis. It begins by explaining the optimization problem, together with different classifications that exist concerning different criteria such as combinatorial versus numerical (based on the type of variables used to describe the optimization problem) or single- versus multi-objective (based on the number of objectives that are optimized). Next, the benchmarking process is explained in more detail, focusing on the four main steps that should be taken with great care when a benchmarking study is performed: (i) identifying the benchmarking objective, (ii) defining the optimization domain (i.e., selection of the algorithms and problems), (iii) defining a fair experimental design, and (iv) statistically analyzing the experimental data. Finally, this part ends with an introduction to statistical analysis, with a special focus on hypothesis testing and guidelines for selecting an appropriate omnibus statistical test for the performance assessment of meta-heuristic stochastic optimization algorithms.

The purpose of the second part is to introduce the weaknesses of classic statistical analyses that are used to compare experimental data obtained from meta-heuristic stochastic optimization algorithms as the motivation for developing the Deep Statistical Comparison, an approach that provides more robust statistical outcomes when there are outliers present in the data or data values are in an ε-neighborhood. This part begins by introducing the Deep Statistical Comparison ranking scheme in a more general form. Next, its application and extensions in single- and multi-objective optimization are presented using examples.

The third part focuses on the implementation and practical usage of Deep Statistical Comparison approaches. It introduces the DSCTool, which is a web-service-based e-Learning tool for statistical comparisons of meta-heuristic stochastic optimization algorithms. The tool helps users with all Deep Statistical Comparison approaches by guiding them on how to input the data (i.e., optimization algorithm results) should be organized and which statistical test is the most appropriate for their benchmarking scenario. This part also provides a practical implementation of all the examples presented in the second part to guarantee reproducible results and teach users how they can use the Deep Statistical Comparison approaches on their own data.

Acknowledgments

We must thank all the friends and colleagues whose comments and criticism helped to improve the book. We would especially like to mention Barbara Koroušić Seljak, Gordana Ispirova, and Gjorgjina Cenikj.

We would also like to thank the Slovenian Research Agency (research core funding No. P2-0098, and the postdoctoral project Mr-BEC No. Z2-1867) and the European Union's Horizon 2020 research and innovation program under grant agreement No. 692286 (SYNERGY for smart multi-objective optimization) for supporting our work.

At the end, this is no real ending. It is just the place where one story ends and another one begins.

Ljubljana, Slovenia   Tome Eftimov
December 2021   Peter Korošec

Contents

Acronyms

AD	Anderson-Darling
API	Application Programming Interface
BBOB	Black-Box Optimization Benchmarking
DSC	Deep Statistical Comparison
eDSC	extended Deep Statistical Comparison
FWER	Family-Wise Error Rate
GUI	Graphical User Interface
IEEE	Institute of Electrical and Electronics Engineers
JSON	JavaScript Object Notation
KS	Kolmogorov-Smirnov
moDSC	multi-objective Deep Statistical Comparison
NP	Non-deterministic Polynomial-time
pDSC	practical Deep Statistical Comparison
REST	Representational state transfer
TSP	Traveling Salesman Problem
XAI	eXplainable Artificial Intelligence

Chapter 1
Introduction

1.1 Motivation

As we move into the era of eXplainable Artificial Intelligence (XAI), a comprehensive comparison of the performance of meta-heuristic stochastic optimization algorithms has become increasingly important to understanding the behavior of algorithms. One of the most common ways to compare the performance of stochastic optimization algorithms is to apply statistical analyses using experimental data. However, there are still caveats that need to be addressed before we can draw relevant and valid conclusions. First, statistical analyses require a knowledgeable user to apply them properly. As the appropriate knowledge is often lacking, this can lead to incorrect conclusions. Next, standard approaches can be influenced by outliers (e.g., poor runs) or some statistically insignificant differences (solutions that are within an ϵ-neighborhood) that exist in the data.

This book provides an overview of the current approaches to analyzing the performance of algorithms, with a special emphasis on the caveats that are often overlooked. We will show how these caveats can be easily circumvented by applying the simple principles that lead to a Deep Statistical Comparison (DSC). The book will not be based on equations, but mainly through the use of examples, from which a deeper understanding of statistics will be achieved. The examples will be based on various comparison scenarios in single-objective and multi-objective optimization algorithms. The book will end with a demonstration of a web-service-based framework called DSCTool for a statistical comparison of stochastic optimization algorithms. The book has an additional seven chapters:

- Chapter 2: Meta-heuristic stochastic optimization
- Chapter 3: Benchmarking theory
- Chapter 4: Introduction to statistical analysis
- Chapter 5: Approaches to statistical comparisons used for stochastic optimization algorithms

© The Author(s), under exclusive license to Springer Nature Switzerland AG 2022
T. Eftimov and P. Korošec, *Deep Statistical Comparison for Meta-heuristic Stochastic Optimization Algorithms*, Natural Computing Series,
https://doi.org/10.1007/978-3-030-96917-2_1

- Chapter 6: Deep Statistical Comparison in single-objective Optimization
- Chapter 7: Deep Statistical Comparison in Multi-objective Optimization
- Chapter 8: DSCTool—a web-service-based e-learning tool.

Next, we provide a brief description of each chapter.

1.2 Meta-heuristic Stochastic Optimization

This chapter provides a short introduction to meta-heuristic stochastic optimization. This will allow the reader to become acquainted with optimization as the subject of a statistical analysis. Besides a brief introduction to optimization, a more detailed description of the combinatorial and numerical (i.e., continuous) optimization families will be provided. In addition, two classes of problems (i.e., single-objective and multi-objective) that are typically solved, will be presented. To conclude, different optimization heuristics and meta-heuristics that are used to solve the desired optimization problems will be introduced.

1.3 Benchmarking Theory

Benchmarking is crucial to making a comparison of optimization algorithms. This chapter provides an introduction to benchmarking theory, which defines all the important steps needed for a valid comparison of different algorithms. Benchmarking consists of four main steps. First, the reasons for benchmarking need to be identified, so an appropriate benchmarking scenario is defined. Second, the optimization domain needs to be set in the form of selecting the appropriate optimization problems and algorithms. Third, the experimental design needs to be defined. Finally, the statistical analysis provides valid results and conclusions.

1.4 Introduction to Statistical Analysis

To allow the reader to become familiar with the statistical terms that will be used in the book, this chapter will provide an introduction to statistical analysis. We will explain the difference between descriptive statistics and inferential statistics. We will also provide the reader with a background in frequentist hypothesis testing, which is the key to making a statistical comparison and always involves two hypotheses: the null and the alternative. Next, we will describe different types of statistical tests (e.g., parametric vs. non-parametric, omnibus vs. post-hoc) and discuss the required conditions that must be met in order to apply them properly. The selection of a statistical test is crucial to the outcome of a study because applying an inappropriate test can lead to the wrong conclusion. This will be followed by a brief explanation of the different statistical scenarios, including a pairwise comparison, multiple comparisons, and multiple comparisons with a control algorithm.

1.5   Approaches to Statistical Comparisons Used for Stochastic Optimization Algorithms

This chapter will provide an overview of the most commonly used approach to making a statistical comparison of meta-heuristic stochastic optimization algorithms, followed by the latest advance, known as the Deep Statistical Comparison approach, which was invented to provide more robust statistical results. Both approaches will be discussed in single- and multiple-problem benchmarking scenarios. The chapter will also introduce the standard DSC ranking scheme in a more general form, which is the basis for defining all its variants, which will be discussed for single- and multi-objective optimization.

1.6   The Deep Statistical Comparison in Single-Objective Optimization

This chapter will introduce the application of the Deep Statistical Comparison approach in single-objective optimization. We will explain the difference between the practical and the statistical significance. For this, two versions of the practical Deep Statistical Comparison (pDSC) approach will be introduced. The approach is focused on testing whether the statistical significance that can be presented in the algorithms' results is also relevant for real-world applications. In addition, the extended Deep Statistical Comparison (eDSC) approach will be introduced, which is used to compare single-objective optimization algorithms concerning the distribution of the obtained solutions in the search space. This approach provides more information about the exploration and the exploitation capabilities of the compared algorithms. Finally, all the introduced DSC variants will be presented, with examples to make the reader familiar with their working principles.

1.7   The Deep Statistical Comparison in Multi-objective Optimization

This chapter will introduce the application of the Deep Statistical Comparison approach in multi-objective optimization. We will show how the standard Deep Statistical Comparison ranking scheme can be used to compare the performance of multi-objective optimization algorithms using a single quality indicator data. The comparison is even more complicated because the selection of quality indicators as performance measures for multi-objective optimization algorithms can be biased to the user's preference. For this purpose, we will introduce three ensemble heuristics (i.e., an average, a hierarchical majority vote, and a data-driven) that can be used to compare the algorithms based on a set of user-defined quality indicators in order to make a more general conclusion. Finally, the multi-objective Deep Statistical Comparison (moDSC) will be introduced. This reduces the user's bias in the selection of the quality indicators and provides more robust statistical results by reducing the loss of information when the high-dimensional data (i.e., the approximation sets) are transformed to one-dimensional data (i.e., quality indicators). Finally, all the introduced DSC variants for multi-objective optimization algorithms will be presented with examples to make the reader familiar with their working principles.

1.8   DSCTool—A Web-Service-Based e-Learning Tool

The last chapter will present examples so that the reader will become familiar with a web-service-based framework called DSCTool for making a statistical comparison of meta-heuristic stochastic optimization algorithms easier. We will describe how the Deep Statistical Comparison web-service-based framework can be used to avoid drawing incorrect conclusions. The examples will include the implementation of comparisons of single-objective and multi-objective optimization algorithms that are included in Chaps. 6 and 7.

Chapter 2
Meta-heuristic Stochastic Optimization

2.1 Optimization

Optimization is a process or methodology in the form of a mathematical procedure, the goal of which is to find the *minimum* or *maximum* of a function. Optimization can be represented as follows. Given a function $f : A \rightarrow \mathbb{R}$ that translates the elements from set A to the real numbers, we search for the elements $x^* \in A$ that return the optimum or best $f(x^*)$ with respect to some criteria. In the case of minimization, we look for such x^* that $f(x^*) \leq f(x)$ for all $x \in A$. While in the case of maximization, $f(x^*) \geq f(x)$ for all $x \in A$. In essence, we have defined the general formulation of an optimization problem.

The function f is called the *objective function*. This function can range from a simple mathematical equation to a complex algorithm that simulates a process. The objective function domain A is called the *problem* or *search space* and its elements *candidate solutions*. A set A can be represented by different types of elements like arrays of numerical, categorical or mixed variables. A can be additionally specified with a set of constraints that need to be satisfied. The elements of A that satisfy the constraints are called *feasible solutions*. The feasible solution that optimizes the objective function is called the *optimal solution* or *global optimal solution*.

When the objective function of the problem is not purely convex or concave, there might be many local optima in the form of local minima or maxima, respectively.

With respect to the variables that define the optimization problem, the optimization is divided into two main problem families: *combinatorial* and *numerical*. Combinatorial optimization deals with optimization problems defined over discrete variables, while numerical optimization deals with optimization problems defined over continuous variables.

The *local minimum* $x' \in A$ is defined as the point for which there exists some $\varepsilon > 0$, so that for all $x \in A$ such that $\|x' - x\| \leq \varepsilon$, the expression $f(x') \leq f(x)$ holds. The distance between two solutions is defined by a distance measure. For example, for real variables this can be an *Euclidean distance*, while for combinatorial

© The Author(s), under exclusive license to Springer Nature Switzerland AG 2022
T. Eftimov and P. Korošec, *Deep Statistical Comparison for Meta-heuristic Stochastic Optimization Algorithms*, Natural Computing Series,
https://doi.org/10.1007/978-3-030-96917-2_2

variables this is defined by *neighborhood relations* (e.g., in a graph this could be an edge that connects two vertices). The *local maximum* is defined analogously, where the relation $\leq$ is replaced by $\geq$.

2.1.1   Combinatorial Optimization

Combinatorial optimization [14] algorithms deal with optimization problems where the set of feasible solutions is *finite* and *discrete* or can be reduced to a *discrete* one, and have a corresponding graphical representation. Even though the search space is finite, it is often still large and needs to be explored efficiently. To this end, combinatorial optimization algorithms tackle such problems by reducing the effective size of the search space by exploring it in some smart way (e.g., by finding and exploring only the promising regions).

A *combinatorial optimization problem* $C = (X, D, \Omega, f, extr)$ can be defined with

- a set of discrete variables $X = \{x_1, \ldots, x_n\}$,
- a set of domains $D = \{D_1, \ldots, D_n\}$ of variables, where the discrete variable x_i is taken from the domain $D_i, i = 1, \ldots, n$,
- a finite set of constraints, Ω, defined over the variables X, which define a space of feasible solutions $S \subseteq X$,
- an objective function $f : D_1 \times \cdots \times D_n \to \mathbb{R}$,
- the extreme $extr$, which is either a min or max.

To solve a combinatorial optimization problem, a feasible solution $s^* \in S$ has to be found, such that an $extr$ value is returned by the objective function f. Therefore, $f(s^*) \leq f(s)$ must be true for $\forall s \in S$ in the case when we are looking for a minimum, or $f(s^*) \geq f(s)$ must be true for $\forall s \in S$ in the case of a search for a maximum. Solution s^* is called the global optimum solution of the problem C. The set of global optimum solutions for a given problem is marked with S^*.

Combinatorial optimization problems can be differentiated according to the *generic problem representation* and *corresponding objective function*, which determines what kind of problem class is being solved (e.g., traveling-salesman problem (TSP) [89], bin-packing problem [77], etc.). Each problem class has actual problem instances, which are defined by the problem size n, discrete domains, and/or constraints (e.g., an actual graph instance for TSP). For this reason, we are not saying that we are solving a problem, but a *problem instance*. This is an important distinction that the reader should remember.

2.1.2   Numerical Optimization

Numerical optimization [83] algorithms deal with a *continuous optimization problem*, where the search space is infinite. Here, the objective is to find n real-valued variables, often called *parameters*. For this reason, numerical optimization is some-times referred to as *multi-parameter optimization*. The goal is to find the values of the parameters that optimize the objective function. The domain of numerical opti-mization is optimization problems where the set of *feasible solutions* is continuous, and the goal is to find the best-possible solution.

A *numerical optimization problem* $N = (P, D, \Omega, f, extr)$ is defined by

- a set of continuous parameters $P = \{p_1, \ldots, p_n\}$,
- a set of domains $D = D_1, \ldots, D_n$ of parameters, where the continuous parameter p_i is taken from the domain D_i, $i = 1, \ldots, n$,
- a finite set of constraints, Ω, defined over the parameters P, which define a space of feasible solutions $S \subseteq P$,
- an objective function $f : D_1 \times \cdots \times D_n \rightarrow \mathbb{R}$,
- the extreme $extr$, which is either a min or max.

To solve a numerical optimization problem, a *feasible solution* $s^* \in S$ has to be found such that an $extr$ value is returned by the objective function f. Therefore, $f(s^*) \leq f(s)$ must be true for $\forall s \in S$ in the case when we are looking for a *minimum*, or $f(s^*) \geq f(s)$ must be true $\forall s \in S$ in the case of a search for a *maximum*. The solution s^* is called the global optimum solution of the problem N. The set of global optimum solutions for a given problem is marked with S^*.

Like combinatorial optimization problems, numerical optimization problems are also differentiated according to the objective function, though here the distinction between the problem classes and the problems themselves is not clearly defined. Since the taxonomy is not defined and used consistently, all of them might belong to the same class, or they could be split into real-world and artificial problem classes, or through some other problem characteristics that would define different problem classes. What is common to all numerical optimization problems is that they all have the same *generic representation* in the form of n real-valued parameters, while the representations vary between classes in combinatorial optimization problems. So, to simplify matters, we will, in this book, consider all of the numerical optimization problems as one class with a distinction between problems being defined purely based on the definition of the objective function. Since the objective function alone defines the problem, the *problem instances* are defined by objective function's:

- translations, shifts, and/or scaling,
- problem size n,
- ranges of continuous domains, and/or
- constraints.

This is an important distinction in comparison to combinatorial optimization that needs to be understood and taken into account.

2.2 Optimization Problem

In addition to the classification of optimization problems according to a set of variables (i.e., combinatorial, numerical), they can be classified with respect to *how many objectives* are being optimized. Typically, we distinguish between problems that have only one objective, called *single-objective problems*, and others that have more than one objective, called *multi-objective problems*. Nowadays, for problems that have more than three objectives, the term *many-objective problems* is used.

2.2.1 Single-Objective Optimization Problem

In Sect. 2.1 we described optimization as being applied to single-objective optimization problems (see Fig. 2.1). We can generalize the single-objective optimization problem using the following mathematical formulation:

- optimize $f(x)$)
- subject to

  - $x_i \in D_i, i = 1, \ldots, n,$
  - $g_j(x) \leq 0, j = 1, \ldots, p,$
  - $h_k(x) = 0, k = 1, \ldots, r,$

where $f(x)$ is the objective function that needs to be minimized (or maximized); x is the vector of the solution variables from domains D; $g_j(x)$ is the jth out of

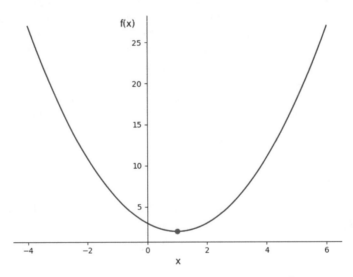

Fig. 2.1 Minimum (red dot) of single-objective function $f(x) = ((x - 1)^2) + 2$

p inequality constraints; and $h_k(x)$ is the kth out of r equality constraints. The constraints define the *feasible region* of the solution space or the *feasible solutions*.

2.2.2 Multi-objective Optimization Problem

We can generalize the multi-objective optimization problem using the following mathematical formulation:

- optimize $F(x) = [f_1(x), \ldots, f_m(x)] \in \mathbb{R}^m, m > 1$
- subject to

    - $x_i \in D_i, i = 1, \ldots, n,$
    - $g_j(x) \leq 0, j = 1, \ldots, p,$
    - $h_k(x) = 0, k = 1, \ldots, r,$

where $F(x)$ is a set of $m > 1$ objective functions $f_i(x)$ that need to be minimized (and/or maximized); x is the vector of solution variables from the domains D; $g_j(x)$ is the jth out of p inequality constraints; and $h_k(x)$ is the kth out of r equality constraints. The constraints define the feasible region of the solution space or the feasible solutions.

Here, we are dealing with multiple objective functions that are often conflicting. This means that an improvement on one objective can only be achieved by a deterioration of the other objectives. As a consequence, in such a case no single optimum solution exists for all objectives, but there are a potentially infinite number of so-called Pareto optimal solutions. A Pareto optimal solution is a feasible solution that cannot be improved in any of the objectives without degrading at least one of the other objectives. Mathematically, for the minimization problem a feasible solution $x_1 \in X$ dominates another feasible solution $x_2 \in X$ if:

- $\forall f_i(x_1) : f_i(x_1) \leq f_i(x_2), i = 1, \ldots, m$
- $\exists f_j(x_1) : f_j(x_1) < f_j(x_2), j = 1, \ldots, m.$

A feasible solution $x \in X$ is called the *Pareto optimal*, if there is no other solution that dominates it. The set of all the Pareto optimal solutions is called the *Pareto set*, while its representation in the objective space $F(x)$ is known as the *Pareto front*. Since this set can be very large or even infinite, the optimization algorithms return a set of non-dominated solutions that are approximating the Pareto front, called an *approximation set* (see Fig. 2.2).

Since the result of a multi-objective optimization is an approximation set, there is no direct way to compare the quality of the results of two optimization algorithms. For this purpose, a large number of *quality measures* have been proposed, where an approximation set is mapped to a *real number*. These mapping functions are called *quality indicators* [90]. The best known are hypervolume [108], generational distance [103], inverse generational distance [103], epsilon [67], spread [18], and generalized spread [18]. This allows for the quality of the results to be measured with respect to different convergence and diversity criteria.

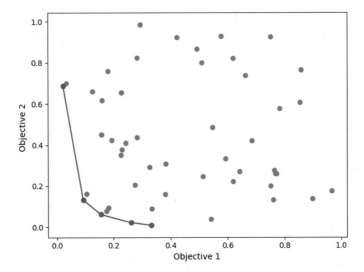

Fig. 2.2 An example of an approximation set (in red) for two objectives that need to be minimized

2.3 Heuristics

There is no universal optimization algorithm that can solve all optimization problems
[107]. Many of the problems arising in real-life applications are NP-hard [1]. In
simple terms, these are the problems for which it is not known whether they can be
solved in polynomial time with respect to the size of the problem. Hence, we usually
solve large instances with approximate methods that return near-optimal solutions
in a relatively short time. Algorithms of this type are called *heuristics*.

2.3.1 Exact Heuristics

Exact heuristics are optimization algorithms designed to guarantee that they will
always find an optimal solution in a finite amount of time. Typically, we are talking about deterministic heuristics, which are heuristics in which no randomness is
involved in the search process. A deterministic heuristic is an optimization algorithm
that will always return the same solution, given the same input. This is because every
step of the algorithm is defined by the previous step and so the underlying optimization process always passes through the same sequence of steps. However, for very
difficult optimization problems (e.g., NP-hard problems) the finite amount of time
can exponentially increase with respect to the dimensionality of the problem, which
makes them impractical for solving such problems.

2.3.2 *Stochastic Heuristics*

Stochastic heuristics generalize deterministic heuristics. Stochastic heuristics are optimization algorithms that generate and use random variables to determine the next steps in the optimization process. Typically, for generating random variables a pseudo-random number generator [63] is used. For this reason, stochastic heuristics do not return the same solution every time using the same input and do not guarantee to find the optimal solution, as the exact heuristics do. Therefore, stochastic heuristics generally return solutions that are worse than optimal. However, stochastic heuristic algorithms usually find good/acceptable solutions in a reasonable amount of time, even for NP-hard problems.

2.4 Meta-heuristics

Meta-heuristic is in essence an upgrade of a heuristic, where a set of algorithmic concepts that can be used to define a heuristic method are applicable to a wider set of different problems. While the majority of heuristic algorithms are very problem specific, meta-heuristics are designed with a goal to solve problems at some higher level. They combine subordinate, more problem-specific procedures—usually heuristics themselves—in what is hoped to be an efficient way. The name meta-heuristic, first introduced by Glover [50], combines the Greek word *meta* ($\mu\varepsilon\tau\alpha$), meaning "beyond", here in the sense of "higher level", and the verb *heuriskein* ($\varepsilon\upsilon\rho\iota\sigma\kappa\varepsilon\iota\nu$), which means "to find". Since there is no commonly accepted definition for the term meta-heuristic, we have gathered some of them below.

> A meta-heuristic is formally defined as an iterative generation process that guides a subordinate heuristic by combining intelligently different concepts for exploring and exploiting the search space; learning strategies are used to structure the information to find near-optimal solutions efficiently.
>
> — Osman and Laporte [84]

> Metaheuristics are typically high-level strategies that guide an underlying, more problem-specific heuristic, to increase their performance. The main aim is to avoid the disadvantages of iterative improvement and, in particular, multiple descent, by allowing the local search to escape from local optima. This is achieved by either allowing worsening moves or generating new starting solutions for the local search in a more "intelligent" way than simply providing random initial solutions. Many of the methods can be interpreted as introducing a bias such that high-quality solutions are produced quickly. This bias can be of various forms and can be cast as a descent bias (based on the objective function), a memory bias (based on previously made decisions) or an experience bias (based on prior performance). Many of the metaheuristic approaches rely on probabilistic decisions made during the search. However, the main difference when compared to a pure random search is that in metaheuristic algorithms the randomness is not used blindly, but in an intelligent, biased form.
>
> — Stützle [99]

A metaheuristic is an iterative master process that guides and modifies the operations of
subordinate heuristics to efficiently produce high-quality solutions. It may manipulate a
complete (or incomplete) single solution or a collection of solutions during each iteration.
The subordinate heuristics may be high-level (or low-level) procedures, or a simple local
search, or just a construction method.

— Voß et al. [104]

A metaheuristic is a set of concepts that can be used to define heuristic methods that can be
applied to a wide set of different problems. In other words, a metaheuristic can be seen as a
general algorithmic framework that can be applied to different optimization problems with
relatively few modifications in order to adapt them to a specific problem.

— Metaheuristics Network [80]

Blum and Roli [10] summarized the fundamental properties of meta-heuristics as
follows:

- Meta-heuristics are strategies that "guide" the search process.
- Their goal is to efficiently explore the search space in order to find (near-)optimal
  solutions.
- Techniques that constitute meta-heuristic algorithms range from simple local-
  search procedures to complex learning processes.
- Meta-heuristic algorithms are approximate and usually non-deterministic.
- They may incorporate mechanisms to avoid getting trapped in confined areas of
  the search space.
- The basic concepts of meta-heuristics permit an abstract-level description.
- Meta-heuristics are not problem-specific.
- Meta-heuristics can make use of domain-specific knowledge in the form of heuris-
  tics that are controlled by the upper-level strategy.
- Today's more advanced meta-heuristics use search experience (embodied in some
  form of memory) to guide the search.

Meta-heuristics are high-level strategies for exploring search spaces using a vari-
ety of methods. However, it is very important that a dynamic balance is maintained
between diversification and intensification. The term diversification generally refers
to the exploration of the search space, whereas intensification refers to the exploitation of the accumulated search experience.

Meta-heuristics are generally applied to problems for which there is no satisfactory
problem-specific algorithm or heuristic, or when it is not practical to implement such
a method.

2.5 Summary

In this chapter:

- We introduced the basic terms used in optimization, with a special focus on the differences between combinatorial and numerical optimization.
- We introduced the classification and main differences between the optimization problems based on how many objectives are being optimized, single- or multi-objective optimization.
- We provided an overview of different heuristics that can be used to perform optimization, with a special focus on meta-heuristics.

Chapter 3
Benchmarking Theory

3.1 Benchmarking

When developing a new optimization algorithm, the most important step is under-standing its behavior and performance. This includes understanding the influence of the algorithm's hyper-parameters and/or various operators that define the behavior of the algorithm in regard to the performance achieved. Since the process that leads to a proper and unbiased understanding can be very time consuming and complex, in recent years a *benchmarking theory* has evolved. It guides the process with some high-level directions to gain the most relevant information using the obtained results from the experiments. The benchmarking theory consists of four main steps:

1. identify the reason for benchmarking,
2. define the optimization domain (problem and algorithm selection),
3. define and execute an experimental design,
4. analyze the experimental results using statistical analyses.

3.2 Objectives of Benchmarking

There can be many reasons why we would like to perform a benchmarking study. A detailed explanation can be found in [5], but here we will only summarize the main reasons to grasp the basic concepts behind them.

- *Visualization and basic assessment* is the most commonly performed published study. The goal is to assess the performance and behavior of different compared algorithms. The comparison can be made on some predefined set of optimiza-tion algorithms, or in some type of competitions where researchers follow the same experimental procedure after which all the algorithms are compared by the competition organizers. To gain a better understanding of an algorithm's behavior,

© The Author(s), under exclusive license to Springer Nature Switzerland AG 2022
T. Eftimov and P. Korošec, *Deep Statistical Comparison for Meta-heuristic Stochastic Optimization Algorithms*, Natural Computing Series,
https://doi.org/10.1007/978-3-030-96917-2_3

different visualization techniques (e.g., plotting convergence curves) are also common. Probably less researched and, arguably, at least as important is the assessment of the optimization problems, as will be presented in the following Sect. 3.3.

- *Sensitivity of performance* is primarily related to understanding an individual algorithm's behavior and performance. This includes testing the algorithm's invariance to the influence of different problem transformations (e.g., translation, scaling, and rotation), understanding the influence of different algorithm configurations (i.e., algorithm's hype-parameters), and identifying algorithm's performance preferences to some type of problems (e.g., high performance on separable problems).
- *Performance extrapolation*'s goal is to identify an algorithm's performance regression, which can be, in different forms, used for successful automated algorithm design, selection, and configuration.
- *Theory-oriented goals* arise from the idea to further theoretically expand the knowledge on different, already achieved results in the form of some theoretical, empirical, or practical results.
- *Benchmarking in algorithm development* is most often used for code validation and general algorithm development, which includes an empirical comparison of new ideas, and their influence on the algorithm's behavior and performance.

Identifying the reason for benchmarking is an important step, since all the subsequent steps in the benchmarking are determined by it. In this book we will not go into the details of each step, but only an overview presentation for the reader to gain a better grasp of what is involved. Understanding the reasons for benchmarking can guide and help users in the proper selection of the optimization problems, optimization algorithms, overall experimental design, and how and which results need to be analyzed to achieve the desired goal of the benchmarking study.

3.3   Problem Selection

Problem selection is one of the most critical steps in benchmarking since the choice of problems and their instances can heavily influence the results. Looking from the perspective of an ideal world, we would like to select problems that cover the whole problem space, which is difficult to define. This brings us to the definition of the problem space, which is typically defined in terms of the *problem features* that describe the problems themselves. These features can be *descriptive* (e.g., separability, multimodality, ill-conditioning, etc.) or *analytical* (e.g., using a set of exploratory landscape-analysis features). Descriptive features are general for the problem, not quantified (e.g., there is no quantification as to what extent a problem is multimodal), and require a prior, in-depth knowledge about the problem. On the other hand, analytical features are problem-specific, are quantifiable, and do not require in-depth knowledge about the problem since they are based on calculations using a sampling strategy. Unfortunately, there is an issue of the identification and selection of relevant features and the sampling strategy that will be used for their calculation. Currently, we do not know which set of features would best represent the problem space and what kind of sampling strategy should be applied [97]. Even though there is no clear definition of the problem space, this does not mean that nothing can be done with regard to problem selection. The issue can be looked at from the other perspective: not selecting the problems that cover the whole problem space, but selecting the problems that are diverse (i.e., they do not share the same/similar features). This can be looked at from the perspective of problem features (as mentioned above) or *performance features*. The performance features are determined by a set of optimization algorithms run on a set of optimization problems. Instead of only analyzing the characteristics of the problem, we are analyzing how algorithms performed (their ranking) on individual problem instances. If the rankings are the same for two problem instances, the instances are considered to be similar and vice versa. From the perspective of benchmarking it is desirable to have the minimum number of problem instances, but enough to make a proper statistical analysis, since this drastically reduces the benchmarking time complexity.

Though each of the reasons presented in the Sect. 3.2 has its own specifics related to the problem selection, we can make a selection for all of them with respect to one of the following two goals:

- Benchmarking with the goal of analyzing an algorithm with respect to a single/some *specific types of problems*. In really specific cases, this might be only a single problem instance (a real-world optimization problem or its simulation) or a set of very similar problems that are considered to have the same/similar features.
- Benchmarking with the goal to estimate a more general performance of the algorithm on a *wider set of problems* (e.g., a set of artificial test functions from a benchmark suite) with different features. As mentioned previously, in the ideal case the set of problems would cover the whole problem space.

Regardless of the selected problems, we must be aware that all the conclusions that will derive from the analyses only hold for the problems on which the benchmarking was performed. This is very important to understand and to present in the results.

As mentioned in Sect. 2.1, optimization problems can be divided into two classes: real-world and artificial. Let us describe the main differences between them.

3.3.1 Real-World Problems

Real-world problems are those encountered in real life. As a consequence, they are usually poorly described and there is little if any a priori knowledge about them. They are often considered to be black-box problems (where we have no knowledge about the problem) or grey-box problems (where we have very limited knowledge about them), which can be used for better algorithm selection and/or customization. Furthermore, the evaluation of such problems is often not exact, since there is no mathematical definition of the problem and the evaluation is made according to a simulation that tries to mimic the desired real-world scenario to the best of its abilities. Lastly, each evaluation of such simulations is often very time consuming.

Due to these above-mentioned deficiencies, it is very hard to solve real-world problems efficiently. Since they are typically black/grey-box it is difficult to select the best optimization algorithm and its configuration for solving them. Due to the time-consuming evaluation, it is hard to acquire any meaningful analytical information (very limited sampling size) and the algorithms have only a small number of evaluations at their disposal to find the best possible solution. Although they are the problems that we would like to solve, because they have a real-life impact, they are often not suitable for benchmarking.

3.3.2   Artificial Problems

Artificial problems are typically hand-made problems in the form of functions (though they can be also automatically generated [101]), for which there is a lot of a-priori knowledge already available. In contrast to real-world problems, they are typically white-box problems (the problem landscape, or even the optimum solution(s) are known) and are quick to evaluate. This means we have a relatively easy/fast calculation of the desired analytical features that can be used for describing them. As such, they are most often used as the problems on which benchmarking is performed. The white-box characteristic allows researchers to select the most appropriate problems for the benchmarking algorithm with respect to the desired goal/features. For this reason, they can provide a much better understanding of why and how algorithms behave with respect to the problem features. Although artificial problems sometimes seem to have no real purpose, this is not the case. In an ideal case, they would have all the features of real-world problems and could be used as a perfect training ground for benchmarking optimization algorithms. In such a case, the selected algorithm instance (according to a problem's characteristics) could then be applied to the specific real-world problem with the best possible performance. This makes it possible to transfer academic knowledge to real-world applications.

3.4   Algorithm Selection

When we are benchmarking a new algorithm, it is also really important to select the appropriate algorithms for comparison. When we are developing an algorithm, it is common to compare the different instances of the algorithm (e.g., different hyper-parameters or different configurations/designs). The reason is that we want to find the algorithm instance that will perform best on the desired/selected problem instance(s). In this phase of the algorithm's development, we are typically not interested in the performances of other algorithms. We might have information about the best results achieved by other state-of-the-art algorithms, but this is just to have a general idea of the performance of the algorithm being developed. When we have identified the best-performing algorithm, then we need to identify the algorithms that need to be taken into consideration for the comparison with the newly developed algorithm. Too often the authors of papers that propose new algorithms select such set of (old) algorithms that the result of comparison indicates that the newly developed algorithm is providing some new performance that was not observed before. But such approach is wrong, and any such research studies should be discarded because this does not show the real value of the proposed algorithm. To make the comparison and also the results of the study valid and relevant, we must take great care when selecting the algorithms for a comparison. The selected algorithms must be state of the art at the time of the study, which requires a thorough investigation. In recent years this was helped to some degree by different competitions organized at different conferences and workshops, from which it is possible to acquire a set of problems and the set of algorithms that participated at the event. Unfortunately, taking the winner(s) of the latest competition as a comparison algorithm does not also mean that this was the best algorithm for solving the problem set. As already shown [95, 96], over the years of competition, the newly proposed algorithms might not be better than the ones which participated in previous years. So selecting the winner(s) of the last competition is a good starting point, but this also needs to be taken into consideration. Also when the desired problem features cannot be found in such competitions, then we must look at the various published papers that deal with optimization problems that have the same or similar features. Since typically such papers do not consist of many algorithms, it is always good to also select the best-performing algorithms from competitions that deal with the same type of problem to get a better understanding of the general performance of the proposed algorithm.

In general, this is often not an easy task, but every effort must be made to ensure that the best algorithms are selected for the purposes of the comparison.

3.5 Experimental Design

An appropriate experimental design is essential for a fair comparison. In order to achieve it, the following items need to be considered.

- All the compared algorithms run on the same computer hardware and software, are implemented in the same programming language, and preferably by the same or a similarly skilled programmer. This is important if we want to compare algorithms based on running time or memory usage. If none of them are relevant for the study, then these are not so important.
- All the compared algorithms use the same data type. Changing the data-type representation from "double" (a 32-bit representation of a real number) to "long double" (a 64-bit representation of a real number) can have a measurable impact on the algorithms' performance. With higher real-number precision a result closer to the optimum can be found. The higher precision can also reduce the influence of the

rounding error in the performed objective-function calculations that can have a significant influence on the process of exploration and exploitation of the problem's search space (e.g., this is particularly true for regions of the search space with really small differences). To reduce the influence of the data representation, an ϵ-neighborhood can be defined (e.g., 10^{-8}, so all the algorithms that reach the optimum (if it is known) within this neighborhood are considered to have found the optimum solution. With this we can greatly improve the quality of measures like success rate and computational accuracy.

- All the compared algorithms' hyper-parameters are tuned in the same way. Though this is often hard to achieve, it is essential for a fair comparison. Often, it is clear that the new algorithm was in some way tuned for the selected benchmark problems, while the compared algorithms were tuned on some other problems or they were not tuned at all (e.g., default values were selected). This gives an unfair advantage to the tuned algorithm (i.e., could only be tuned by running some experiments to find generally good hyper-parameters or using some specialized tuning tools like iRace [75], SMAC [62]) in comparison to others that did not receive the same configuration treatment.
- For reproducibility or replicability it is important that the initial seeds for all the experiments are stored. The code must be programmed in such a way that it enables reproducibility or replicability of the experiments by applying the same seed. If the source code is not provided, it is also important to state which programming language and which libraries were used in the experiment. Without this information, replicability is almost impossible, so for this reason, nowadays providing the source code (including experimental data) is essential. As a good practice, the use of some specialized tools (e.g., SnakeMake [71]) for improved reproducibility is recommended.
- If we are dealing with stochastic optimization algorithms, each algorithm must be run many times on the same problem instance, depending on the statistical test that will be utilized.
- With multiple runs we can successfully cover the randomness that is found in different optimization algorithms. Nevertheless, it is still desirable that all algorithms start from the same initial solutions when comparing similar algorithms (e.g., population based or trajectory based). It is well known that different initial solutions can drastically influence the performance of the algorithms (i.e., convergence and quality of the final solution).
- All the compared algorithms should have the same ending condition (e.g., number of problem-solution evaluations achieved close to optimum). This is really essential for a fair comparison.
- When comparing different types of optimization algorithms (e.g., standard and those using surrogate models) it is important that the algorithm type is taken into account when defining the performance measure, since the time complexity of different types of optimization algorithms can be different. This is less important in cases when the evaluation of the solution is time consuming (e.g., one simulation takes hours or days) in comparison to when the evaluation is much faster than one iteration of the optimization algorithm.

- If the goal of benchmarking is to estimate the generalized performance of the optimization algorithm, a wide range of problems should be selected for comparison. Ideally, all the problems would have different features that would cover the whole problem space. Currently, this is not yet possible and well understood, so usually we take a set of artificial test functions from some benchmark suite(s). In cases when we are looking for the best-performing optimization algorithm for some specific problem instance (e.g., real-world problem), then experiments can only be run on a single or some small set of test functions that are known to have desirable features (e.g., found in a real-world problem).

When benchmarking we are often concerned with a run-time analysis that studies the time needed to reach the optimum solution. Most commonly used approaches are known as *fixed-budget analysis* and *fixed-target analysis*. Fixed-budget analysis limits the number of solution evaluations (i.e., budget) to some predetermined value within which the optimum solution should be found. On the other hand, a fixed-target analysis defines some quality threshold (i.e., target), usually the optimum value or some ϵ-neighborhood, when it is known, and studies the time (e.g., the number of evaluations) that are needed to reach the targeted quality. To speed up the benchmarking process, additional constraints are often put in place (e.g., stop the algorithm when it converged or there was no improvement for some period).

3.6 Statistical Analysis

After running all the experiments according to the selected experimental design the last step is to analyze the results. This must be performed appropriately so that valid conclusions of a benchmarking study can be made. This is the main topic of this book and the following chapters concentrate on all the important aspects of the statistical analysis and provide detailed information about state-of-the-art approaches to different benchmarking scenarios [32].

3.7 Summary

In this chapter:

- We explained the objectives and main steps required to perform a benchmarking study for a performance assessment of meta-heuristic stochastic optimization algorithms.
- We introduced how the problem instances and the algorithms involved in the benchmarking study should be selected to make a fairer statistical comparison.
- We provided details about the experimental design that is crucial for a fair statistical comparison.

Chapter 4
Introduction to Statistical Analysis

4.1 Statistical Analysis

Statistics can be interpreted in various ways. According to the most used definition, statistics is a branch of mathematics that is related to the collection, analysis, interpretation, and presentation of data [98]. In general, it is used for analyzing and summarizing a given data set, which can consist of experimental data or data from real-life studies. The idea is that by analyzing a sample of a larger population, interpretations of the population can be developed using the outcomes from the sample.

Classical statistics distinguishes between two types of analyses: *descriptive statistics* and *inferential statistics*. To explain them, we introduce some of the basic terms used in statistics, followed by a special focus on *hypothesis testing*.

A *population* is a set of similar items or events that are of interest for an experiment. A *representative data sample* is a set of data collected and/or selected from a statistical population using some defined procedure [98]. In some cases, it can happen that the selection of the sample to be analyzed involves a selection bias defined in the sampling procedure. One way to obtain a representative sample is to use random sampling, which involves a random selection of the instances from the target population and minimizes the bias. If the sample is not representative, the result is known as sampling error.

Once a representative data sample is selected, it becomes important to know what type of data has been collected. There are two types of data: *qualitative* (categorical) and *quantitative* (numeric). Each type is split into two sub-types: *ordinal* and *nominal* for qualitative data, and *discrete* and *continuous* for quantitative data. *Ordinal* data is categorical, where the values have natural, ordered categories and the distances between the categories are not known. *Nominal* data is also categorical data, but in this case there is no natural order between the categories. *Discrete* data can only take on a finite number of numerical data, while *continuous* data can take on an infinite number of values.

© The Author(s), under exclusive license to Springer Nature Switzerland AG 2022   23
T. Eftimov and P. Korošec, *Deep Statistical Comparison for Meta-heuristic Stochastic Optimization Algorithms*, Natural Computing Series,
https://doi.org/10.1007/978-3-030-96917-2_4

After the data is collected, the next step is to apply statistics using the representative data sample. As we said before, there are two different statistical analyses that can be applied: descriptive statistics and inferential statistics [105]. *Descriptive* statistics summarizes the data from a sample using measures, while *inferential* statistics draws conclusions from the data set that is subject to random variation.

4.2   Exploratory Data Analysis

Descriptive statistics includes a *distribution* of a single variable, measures of *central tendency*, which include the *mean*, *median*, and *mode*, and measures of *variability*, which include *standard deviation*, *variance*, and the *minimum* and *maximum value*. The *distribution* is a summary of the frequency of individual values or a range of values for a variable. It represents every value of the variable and the number of times the value appears in the data sample. The central tendency of a distribution is an estimate of the "center" of a distribution. The *mean* (average) is the most commonly used measure for the central tendency and to compute it we sum all the values in our data sample and then divide the sum by the sample size (number of values in the data sample). The *median* is the value that separates the upper half of a data sample from the lower half, or it can be assumed to be the "middle" value of an ordered data sample. The *mode* is the value in a data sample that appears with the highest frequency. The variability of a distribution refers to the spread of the data values around the central tendency. The *minimum* and the *maximum* values are the minimum and the maximum values that appear in the data sample. The *standard deviation* is used to quantify the amount of variation or dispersion of a set of data values. A low standard deviation indicates that the data values are close to the mean, while a high standard deviation indicates that the data values are spread over a wider range of values. The *variance* is the square of the standard deviation.

Inferential statistics arise because sampling naturally incurs a sampling error and thus a sample is not expected to represent the population perfectly. The methods for inferential statistics include an estimation of the distribution parameters and the testing of statistical hypotheses.

When we talk about statistical analysis, it is also important to distinguish between *univariate*, *bivariate*, and *multivariate* statistical analyses. Univariate analysis works when the data sample only has one variable. It does not deal with relationships and its major purpose is to describe the data. Bivariate analysis involves analyzing two variables to find a relationship between them. Bivariate analysis is a special case of multivariate analysis when multiple relations between multiple variables are analyzed.

4.3 Hypothesis Testing

One of the most commonly used approaches for testing the relationships between two or more data samples is to use hypothesis testing [106]. Hypothesis testing is a procedure in which sample(s) data is used to evaluate a hypothesis. Hypothesis testing, also called significance testing, is one method that can be used to test a hypothesis about a parameter in a population, using data measured in a data sample, or about the relationship between two or more populations, using data measured in data samples.

The method starts by defining two hypotheses, the *null hypothesis* H_0 and the *alternative hypothesis* H_A. The null hypothesis is a statement that there is no difference or no effect and the alternative hypothesis is a statement that directly contradicts the null hypothesis by indicating the presence of a difference or an effect. This step is crucial, because misstating the hypothesis will disrupt the rest of the process. We can define the hypothesis about a single population or about the relationship between two or more populations, e.g., between either the means or variances of a single, two, or multiple distributions or even about the whole distribution.

The next step is to select an appropriate *test statistic* T, which is a measurable function of a random sample that allows researchers to determine the likelihood of obtaining the outcomes if the null hypothesis is true. The level of significance α, also called the *significance level*, which is the probability threshold below which the null hypothesis will be rejected, also needs to be selected.

The last step is to make a decision either to reject the null hypothesis in favor of the alternative or not to reject it. This decision can be made using two different approaches. In the first, all the possible values of the test statistic for which the null hypothesis is rejected, also called the *critical region*, are calculated using the distribution of the test statistic and the probability of the critical region, which is the level of significance α. Then the observed value of the test statistic T_{obs} is calculated according to the observations from the data sample. If the observed value of the test statistic lies within this critical region, the null hypothesis is rejected, and if not, it is not rejected. In the second approach, instead of defining the critical region, a *p-value*, i.e., the probability of obtaining the sample outcome, given that the null hypothesis is true, is calculated. The null hypothesis is rejected, if the p-value is less than the significance level (typically 0.05 or 0.1), and if not, it is not rejected.

In reality, the null hypothesis can be either true or false, and the result of a statistical test can be that the null hypothesis is either rejected or not rejected. When performing hypothesis testing, two types of errors can occur: *Type I* and *Type II*. A *Type I* error occurs if we reject the null hypothesis when it is true ($\alpha = P(\text{Type I Error})$). A *Type II* error occurs if we fail to reject the null hypothesis when the alternative hypothesis is true ($\beta = P(\text{Type II Error})$). The probability of a type-I error is the level of significance (α). So, before the study we usually assign it a small value (e.g., 0.05, 0.01) because researchers do not want to have a type-I error. The probability of a Type-II error is referred to as β. and it is related to the *power* of a statistical test.

The power is the probability that the test will reject a false null hypothesis, or *power* = $1 - \beta$ [73].

Power analysis is an important aspect in any experimental design [43]. It allows researchers to determine the sample size required to detect an effect of a given size with a given degree of confidence. It allows researchers to find the probability of detecting an effect of a given size with a given significance level, under sample-size constraints. If the probability is very low, then researchers can change the sample size of the experiment.

4.3.1   Parametric Versus Non-parametric Statistical Tests

Two types of statistical tests exist: *parametric* and *non-parametric*. To select an appropriate statistical test, and to choose between a parametric and a non-parametric test [48], the first step is to check the assumptions of the parametric tests, also called the required conditions for the safe use of parametric tests. These conditions include checking for *independence, normality of the data*, and *homoscedasticity of the variances*.

Statistically, two events are *independent* if the occurrence of one does not influence the probability of the other occurring. *Normality* indicates that the data is normally distributed, which we can check using statistical tests such as the *Kolmogorov-Smirnov (KS)* [74], *Anderson-Darling (AD)* [86], *Shapiro-Wilk* [93], and *D'Agostino-Pearson* tests [85]. The result from a statistical test can be graphically proven using *histograms* or *Q-Q plots (quantile-quantile)*. In probability, quantiles are points dividing the range of a probability distribution into intervals with equal probabilities.

The *homoscedasticity of variances* indicates the hypothesis of the equality of variances (homogeneity of variances). *Levene's* test [92] can be used to check the homoscedasticity of the variances. If the data does not satisfy the required conditions for the safe use of parametric tests, then the tests can result in incorrect conclusions, and it is better to use an analogous non-parametric test. In general, a non-parametric test is less restrictive than a parametric test, but it is also less powerful than a parametric test, when the required conditions for the safe use of a parametric test are satisfied [48].

In addition to the conditions for the safe use of parametric tests, other parameters that are also related to the selection of an appropriate statistical test are the number of data samples that need to be compared (two or more) and whether the data samples are *paired* or *unpaired*. *Paired* samples (also called dependent samples) are samples in which natural or matched couplings occur. So in the data sample, each data value in one sample is uniquely paired to a data value in the second sample. Examples of paired samples are found when the benchmarking involves a set of multiple optimization problems. The choice between paired and unpaired samples depends on the experimental design. Researchers need to be aware of this when designing their experiments.

4.3.2 Statistical Scenarios

Table 4.1 presents different omnibus statistical tests classified according to the conditions that must be met.

The *t-test* is a parametric statistical test that can be used to determine whether the mean values of two unpaired data sets are significantly different from the other [15]. If the required conditions for the safe use of the parametric test are not satisfied, an alternative non-parametric version of the t-test is *Mann-Whitney U test* [76].

The *paired t-test* [60] is a parametric statistical test that can be used for a pairwise comparison. It is used to test whether the difference between two data sets over multiple problems is non-random. It checks whether the average difference over multiple problems is significantly different from zero. To use this test, the required conditions for the safe use of the parametric tests need to be satisfied. The *Wilcoxon signed-rank test* [72] is a non-parametric alternative to the *paired t-test*, which ranks the differences between the two data sets for each problem, ignoring the signs, and compares the ranks for the positive and negative differences. When the required conditions for the safe use of the parametric tests are satisfied, the *Wilcoxon signed-rank test* is less powerful than the *paired t-test*, otherwise it must be used over the *paired t-test*.

The *One-way ANOVA* is a parametric test and is used to determine whether there are any statistically significant differences between the means of three or more independent (unrelated) groups [55]. An alternative non-parametric version is a *Kruskal-Wallis test* [79].

The *repeated-measures ANOVA* [49] is a parametric statistical test that can be used to test the differences between more than two related samples. The null hypothesis is that all the data sets are the same and the observed differences are random. If the null hypothesis is rejected, a post-hoc test can be used to determine which data sets differ. In the case of *ANOVA*, the *Tukey test* and the *Dunnett test* [20] are used as post-hoc tests. The first is for comparing all the data sets with each other, and the second is used for comparisons of all the data sets with one control data set.

Table 4.1 Classification of different statistical tests

	Two data samples	More than two data samples
Parametric	t-test (unpaired) paired t-test (paired)	One-way ANOVA (unpaired) Repeated-measures ANOVA (paired)
Non-parametric	Mann-Whitney U (unpaired) Wilcoxon signed rank(paired)	Kruskal-Wallis (unpaired) Friedman, Friedman aligned, Iman-Davenport (paired)

The *Friedman test* [48] is a non-parametric alternative to the *repeated-measures ANOVA*. It ranks the data sets for each problem separately and, in the case of ties, average ranks are assigned. The *Friedman aligned-rank test* [48] is a modification of the *Friedman test*. In the *Friedman test*, the ranking scheme allows for intra-problem comparisons only, since inter-problem comparisons are not meaningful. But in cases when the number of data sets for comparison is small, comparability among the problems is desirable and in such instances the *Friedman aligned-rank test* is preferred. The *Iman-Davenport test* [48] is also a modification of the *Friedman test*, in which Friedman's statistic is modified to avoid an undesirable conservative effect of Friedman's statistic.

If the null hypothesis in the *Friedman test*, *Friedman aligned-rank test*, or *Iman-Davenport test*, is rejected, a post-hoc test needs to be used. There are different post-hoc tests that can be used in this case, such as *Nemenyi test* [81], *Bonferroni-Dunn test* [22], *Holm procedure* [59], *Hochberg procedure* [58], and *Bergmann-Hommel's procedure* [8, 47]. These are some of the best known post-hoc tests, and a lot of literature for each of them exists, so we are not going to explain them in more detail.

There are also statistical tests to compare the whole distributions of one-dimensional data such as *two-sample Kolmogorov-Smirnov (KS) test* [7], *two-sample Anderson-Darling (AD) test* [87], and the *Kruskal-Wallis test* [79].

The *two-sample KS test* is a non-parametric test of the equality of continuous, one-dimensional probability distributions that can be used to compare two samples. The two-sample KS test is one of the most useful and general non-parametric methods for comparing two samples, as it is sensitive to differences in both the location and shape of the empirical cumulative distribution functions of the two samples.

The *two-sample AD test* is a non-parametric test that is used for comparing one-dimensional distributions. It is used for the same purpose as the two-sample KS test, but it has the advantage of being a more powerful test. In addition, it is especially sensitive to differences in the tails of distributions and can detect very small differences between the distributions.

The *Kruskal-Wallis test* is a non-parametric test for determining whether data samples come from the same distribution. It is a non-parametric alternative to the *One-way ANOVA* and is used for comparing two or more independent samples of equal or different sample size. The problem with using the *Kruskal-Wallis test* is the misinterpretation of the null hypothesis. The null hypothesis of this test is that the average ranks of the groups are the same. Sometimes, we can see that the null hypothesis is given as "*The samples come from populations with the same distribution.*", which is correct when the samples come from the same distribution and the test will show no difference. However, this result can be misleading because only some kinds of difference in the distribution will be detected by the test. For example, if two populations have symmetrical distributions with the same average value, but one is much wider than the other, their distributions are different but the *Kruskal-Wallis test* will not detect any difference between them. Also, the null hypothesis of the *Kruskal-Wallis test* is often said to be that the medians of the groups are the same, but this is only true if we assume that the shapes of the distributions in each group are the same. If the distributions are different, the null hypothesis will be rejected even when the medians are the same.

Using the statistical tests presented above, we can perform three different statistical scenarios when benchmarking stochastic optimization algorithms:

- Pairwise comparison—when the performances of two algorithms are compared. In this scenario, the statistical tests when two data samples are involved should be utilized.
- Multiple comparisons among the algorithms—when more than two algorithms are involved in the comparisons and if there is a difference between their performances, all versus all pairwise comparisons should be calculated. In this scenario, an omnibus statistical test for more than two algorithms should be selected (e.g., Friedman test in paired scenario). Further, if the null hypothesis is rejected, a set of post-hoc procedures (e.g., Nemenyi, Holm, and Shaffer) can be used to perform all versus all pairwise comparisons.
- Multiple comparisons with a control algorithm—when the performance of a newly developed algorithm should be compared with the performances of the state-of-the-art algorithms. Here, more than two algorithms are involved in the comparison. In this scenario, an omnibus statistical test for more than two algorithms should be selected (e.g., the Friedman test in a paired scenario). Further, a set of post-hoc procedures can be utilized to compare the performance of the control method with the other methods (e.g., Bonferroni-Dunn, Holland, Hochberg, and Holm). Another way of performing this is to make multiple pairwise tests (e.g., the Wilcoxon test). However, since we have multiple independent pairwise comparisons, the true statistical significance of combining the p-values from all independent pairwise comparisons must be calculated.

4.4 Benchmarking Scenarios

In a typical scenario in any domain, statistical analysis can be conducted in two ways: a single-problem analysis and a multiple-problem analysis. A single-problem analysis is a scenario where the data derives from multiple independent measurements on one problem. For example, the performance of two different algorithms can be compared using a single problem (e.g., sphere function). In this scenario, each algorithm has multiple runs that are performed on that problem due to its stochastic nature. The multiple-problem scenario is when we compare the performance of the algorithms over multiple problems or we have paired samples. These two scenarios are applicable in any domain and good examples for comparing the behavior of stochastic optimization algorithms are presented in [20, 46–48].

4.5   Guidelines to Select a Relevant Omnibus Statistical Test

Using the above-presented information about the hypothesis testing and different omnibus statistical tests, Fig. 4.1 presents a pipeline for selecting an appropriate statistical test when stochastic optimization algorithms are benchmarked.

The selection of an appropriate omnibus statistical test should follow the information presented in our experimental data. Instead of taking an omnibus statistical test that was utilized in a similar study, the above-presented pipeline should be considered. It can happen that the omnibus statistical test utilized in a similar study is not an appropriate one for our study, since the conditions for using it are not met.

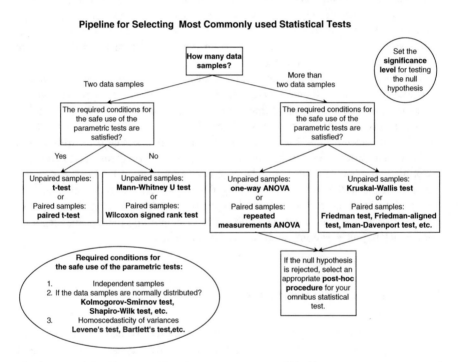

Fig. 4.1   A pipeline for the selection of an appropriate statistical test

4.6 Summary

In this chapter,

- We introduced the basic statistical terms required to understand the book's content.
- We explained the hypothesis testing and its relation to the performance assessment.
- We introduced different statistical and benchmarking scenarios that can be found in a performance assessment of meta-heuristic stochastic optimization algorithms.
- We presented guidelines for selecting a relevant omnibus statistical test.

Chapter 5
Approaches to Statistical Comparisons Used for Stochastic Optimization Algorithms

5.1 Most Commonly Used Approach

When conducting a performance assessment of meta-heuristic stochastic optimization algorithms, the most crucial task is to select an appropriate statistical analysis with which the statistical significance presented in the results will be tested. Based on the statistical outcome of such an analysis, we can decide whether the newly developed algorithm is either statistically significant or not in terms of its performance against the state-of-the-art algorithms.

To compare the results obtained from running meta-heuristic stochastic optimization algorithms, Derrac et al. [21] presented a practical tutorial on how to use nonparametric tests to compare the results. The tutorial follows an already-published tutorial by Demšar [20] presented for a statistical comparison of classifiers over multiple data sets. This approach is the *most commonly used approach* for a statistical comparison of meta-heuristic stochastic optimization algorithms.

5.1.1 Single-Problem Scenario

To compare the results of the algorithms in a single-problem scenario, the multiple runs of the algorithms performed on that single instance of an optimization problem should be analyzed with an appropriate statistical test. In most cases, this scenario involves statistical comparisons made in an unpaired setting, since the order in which the multiple runs are obtained from the algorithms are independent (i.e., the algorithms are run independently, one against the others). However, if the runs of the algorithms are linked, then a paired setting should be investigated. The guidelines on how to select an appropriate statistical test are presented in Sect. 4.5.

© The Author(s), under exclusive license to Springer Nature Switzerland AG 2022
T. Eftimov and P. Korošec, *Deep Statistical Comparison for Meta-heuristic Stochastic Optimization Algorithms*, Natural Computing Series,
https://doi.org/10.1007/978-3-030-96917-2_5

5.1.2   Multiple-problem Scenario

In a multiple-problem scenario, a representative value for each algorithm on each instance of a benchmark problem should be selected from the multiple runs already made. Due to the fact that those algorithms are stochastic in nature, there is no guarantee that the result will be the same for every run. For this purpose, Derrac et al. [21] suggested that the mean value from the multiple runs obtained with an algorithm on a given benchmark problem can be used as an unbiased estimator that should be further involved as a representative value in the statistical comparison. In addition, since the mean value can be affected by outliers presented in the data, the median from the multiple runs can also be used as a representative value. The median can be assumed to be a more robust statistic.

The mean or the median value can be used to represent each algorithm involved in the comparison for each benchmark problem, while the transformed data should be further analyzed with an appropriate omnibus statistical test (see Sect. 4.5).

5.2   Deep Statistical Comparison Approach

We have proposed an approach, called the *Deep Statistical Comparison (DSC)*, to provide more robust statistical data that will be involved when comparing the performance of the meta-heuristic stochastic optimization algorithms [37]. To show the main difference with the most commonly used approach, let us assume two different examples presented below [40].

First Example

Let us assume that two meta-heuristic stochastic optimization algorithms are used to optimize a parabola problem in N dimensions (i.e., $y = \sum_{i=1}^{N} x_i^2$). Each algorithm is run 10 times on the parabola instance and 10 different values are collected. Let us assume that the first algorithm obtained is $\{0, 0, 0, 0, 0, 0, 0, 0, 0, 10\}$ and the second algorithm obtained is $\{0, 1, 0, 1, 0, 1, 0, 1, 0, 1\}$.

Calculating the mean value as a representative value from the multiple runs, the first algorithm has $mean_1 = 1.00$ and the second algorithm has $mean_2 = 0.50$. Assuming that our problem is a minimization problem, it follows that a smaller mean value means better performance. Comparing the algorithms using their mean values, the first algorithm will be ranked as the second, and the second algorithm will be ranked as the first. However, looking at the raw-data values, it follows that the first algorithm has only one poor run that affects the calculation of the mean value. Consequently, we should be aware that outliers exist in the data and can affect the statistical outcome. This example points to an issue that can happen in a single-problem analysis. However, such issue can further affect the test statistic that is used for the multiple-problem analysis.

Second Example

Let us assume that two meta-heuristic stochastic optimization algorithms are used to optimize a parabola problem in d dimensions (i.e., $y = \sum_{i=1}^{d} x_i^2$). For both algorithms 100 runs are performed, and the obtained results are normally distributed with $\mathcal{N}(\mu = 10^{-7}, \sigma = 10^{-5})$.

By randomly sampling from the above distribution, we can generate two samples with a sample size of 100, which can be assumed to be the obtained results by the algorithms. Since the first example indicates that the mean value can be affected by the outliers, let us use the median as a representative value from the multiple runs. When performing a simulation, let us assume that the median values from the multiple runs are $median_1 = 1.18 \times 10^{-6}$ and $median_2 = 5.55 \times 10^{-7}$ for both algorithms, respectively. Since a lower value is preferred, ranking the algorithms based on their median values, the first algorithm will be ranked as the second, and the second algorithm as the first. However, when comparing the median values, it is obvious that they are in an ϵ-neighborhood, $|median_1 - median_2| < \epsilon$. So, the question that arises here is, do we need to rank the algorithms differently, when they have the same distribution of the results that are obtained and their median values are in a small ϵ-neighborhood?

In addition, it can happen that there is no statistical significance between the distributions of the results and their mean/median values are in a small ϵ-neighborhood (see Fig. 5.1a). In such a scenario, the algorithms should be ranked the same. However, it can also happen that there is a statistical significance between the distributions of the obtained results and their mean/median values are in a small ϵ-neighborhood (see Fig. 5.1b). Here, the algorithms should be ranked differently, since their distributions differ.

To reduce the effect from outliers and a small ϵ-neighborhood that is presented between data values [34], the *Deep Statistical Comparison ranking scheme* has been proposed [39], which provides a more robust transformation of the raw data before it is analyzed with an appropriate statistical test.

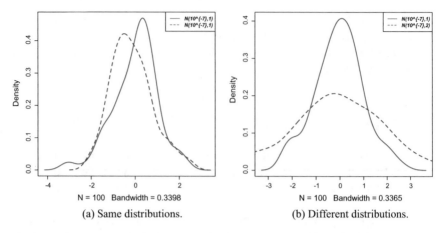

(a) Same distributions.   (b) Different distributions.

Fig. 5.1 Normally distributed data sets from the same and different distributions

5.2.1   Deep Statistical Comparison Ranking Scheme for One-Dimensional Data

Let m be the number of algorithms whose performance should be compared on k benchmark optimization problem instances. Performing the experiment, each algorithm is run n times on the same optimization problem.

The DSC ranking scheme is based on comparing the whole distributions of the one-dimensional data (i.e., obtained solution values), instead of using either the mean or the median value as a representative value. In order to do this, a two-sample test for comparing distributions should be selected and executed with a previously set significance level, α. Using the selected statistical test, all $m \cdot (m - 1)/2$ pairwise comparisons should be performed, and the obtained p-values should be organized into a $m \times m$ matrix, M_i, defined as follows:

$$M_i[p, q] = \begin{cases} p_{\text{value}}, & p \neq q \\ 1, & p = q, \end{cases} \tag{5.1}$$

where p and q are different algorithms, $p, q = 1, \ldots, m$, and i is the problem instance for which the algorithms are compared.

By calculating the p-values matrix M_i, multiple independent pairwise comparisons are performed. In such a scenario, the *family-wise error rate (FWER)* should be controlled. The FWER is the probability of making one or more false discoveries, or type-I errors, among all the hypotheses. When fewer algorithms are compared, the influence of the FWER on the multiple comparisons might not be large. However, when the number of compared algorithms increases, the FWER can increase dramatically. To control it, the obtained p-values should be corrected. For this purpose, several correction methods exist [48], the simplest one being the Bonferroni correction. The idea behind this correction is to test each individual pairwise hypothesis with a significance level of $\frac{1}{C_m^2}$ times the desired significance level, α. The $C_m^2 = \frac{m \cdot (m-1)}{2}$ is the number of pairwise comparisons involved in the calculation of the matrix M_i. By correcting the p-values, the matrix M_i is undergoing a binarization and is presented in the following form:

$$M_i'[p, q] = \begin{cases} 1, & M_i[p, q] \geq \alpha/C_m^2. \\ 0, & M_i[p, q] < \alpha/C_m^2. \end{cases} \tag{5.2}$$

When the element $M_i'[p, q]$ is 1, it follows that the null hypothesis that the two data samples obtained with the pth and qth algorithm come from the same distribution is not rejected. If the element $M_i'[p, q]$ is 0, it follows that the null hypothesis is rejected, so the two data samples come from different distributions.

The matrix M_i' is always symmetrical, $M_i = M_i^T$, and reflexive. The key part for the DSC ranking scheme is to check the transitivity of the matrix, M_i', since the ranking is made in accordance with it. For this purpose, the matrix $M_i'^2$ is calculated. If M_i' has one in each position for which $M_i'^2$ has a non-zero element the transitivity is satisfied, otherwise it is not. The DSC ranking scheme further ranks the algorithms according to the transitivity property.

If the transitivity is satisfied, the set of compared algorithms is split into w disjoint sets of algorithms, Φ_j, $j = 1, \ldots, w$. Within each set, the algorithms' results are coming from the same distribution (i.e., they should be ranked the same), while across different sets the results are coming from different distributions (i.e., they should be ranked as different).

Furthermore, the matrix W_i should be defined, which is a $w \times 2$ matrix involved in the ranking scheme:

$$W_i[j, x] = \begin{cases} \text{mean}(\Phi_j[h]), & x = 1, \\ |\Phi_j|, & x = 2, \end{cases} \tag{5.3}$$

where h is the number ceiled to the nearest integer of a number obtained by the uniform distribution of a random variable $Y \sim U(1, |\Phi_j|)$. Since the algorithms that belong to the same set have the same distributions of data, one algorithm should be randomly selected from each data set as a representative one. Furthermore, its mean value should be involved in the ranking process. The second column provides the number of algorithms that belong to each set. To perform the ranking, the matrix W_i is reordered according to the first column (i.e., the mean values) sorted in ascending order.

Let c be a $w \times 1$ vector that corresponds to the second column of the matrix W_i, which consists of the number of algorithms that belong to the same set. The rankings of the sets should be organized into a $w \times 1$ vector $rank_i$. The set with the lowest mean value should be ranked as

$$rank_i[1] = \sum_{r=1}^{c[1]} r/c[1].$$ (5.4)

For the remaining sets, the rank is defined as

$$rank_i[j] = \sum_{r=c[j-1]+1}^{c[j-1]+c[j]} r/c[j].$$ (5.5)

Next, each algorithm obtains its rankings according to the set to which it belongs.

If the transitivity is not satisfied, the ranking is performed according to the fractional ranking scheme, where a lower value is preferable. This is similar to the most commonly used approach with the advantage that the distributions have already been checked and there is no influence of the outliers or the small ϵ-neighborhood presented in the data values.

5.2.2   Single-Problem Analysis

The DSC ranking scheme works with the input data collected from running the algorithms on a single optimization problem. The obtained DSC rankings contain the information about the statistical significance between the compared algorithms on that single instance of a problem.

5.2.3   Multiple-problem Analysis

In a multiple-problem analysis, where the analysis should be performed using all k instances of the benchmark problem, the DSC ranking scheme should be applied to each instance of the benchmark problem, separately. By using the ranking scheme a $k \times m$ matrix, $Rank$, should be created. The ith row of the matrix corresponds to the DSC algorithms' rankings obtained using the data samples from the ith problem instance. Next, this matrix is used as input data for an appropriate omnibus statistical test in order to draw a general conclusion about the comparison.

5.3   Summary

In this chapter,

- We presented the most commonly used approach for making a statistical comparison of meta-heuristic stochastic optimization algorithms based on using either the mean or median value of the obtained solution values from multiple runs of an algorithm on a problem.

- We pointed out the weaknesses of using the most commonly used approach when outliers are presented in the experimental data or data values are in an ϵ neighborhood.
- We introduced the Deep Statistical Comparison approach that ranks the experimental data based on its whole distribution instead of selecting a single descriptive statistic such as the mean or median value to represent it.

Chapter 6
Deep Statistical Comparison in Single-Objective Optimization

6.1 Statistical Significance

Statistical significance is especially important in Science. Making experimental analysis research publishable requires a statistical analysis that shows the significance of the experimental outcomes. When working on a newly developed algorithm, it is important to compare its performance with the performances achieved by the state-of-the-art algorithms developed for the same purpose. To show the requirement for developing a new algorithm and the investment in the time to do it, at the end we must show that the newly developed algorithm has a statistically significant better performance than the performances of the other already-existing algorithms.

6.1.1 Deep Statistical Comparison Ranking Scheme

The DSC ranking scheme introduced in Sect. 5.2.1 is focused on testing the presence of a statistical significance in the data. Here, we are not going to reintroduce the DSC ranking scheme; however, we are going to provide more details about how the ranking scheme can be applied. The DSC ranking scheme is developed to work with data that is coming from a single-problem scenario. So, the ranking scheme should be applied to every individual optimization problem separately.

The main idea behind the ranking scheme is to compare distributions of the performance data that are obtained by the algorithms being compared. In single-objective optimization, the solution value is one-dimensional data. To apply the DSC ranking scheme, first a statistical test for comparing the distributions of one-dimensional data should be selected. For this purpose, three tests can be used: the *Kruskal-Wallis test*, the *two-sample Kolmogornov-Smirnov (KS) test*, and the *two-sample Anderson-Darling (AD) test*. However, recalling the explanations in Sect. 4.3.2, only two-sample KS and AD are used in the DSC rankings scheme. To reduce the FWER

© The Author(s), under exclusive license to Springer Nature Switzerland AG 2022
T. Eftimov and P. Korošec, *Deep Statistical Comparison for Meta-heuristic Stochastic Optimization Algorithms*, Natural Computing Series,
https://doi.org/10.1007/978-3-030-96917-2_6

the *Bonferroni correction* [46] is used to correct the obtained p-values. We should mention here that the Bonferroni correction is used as a part of the DSC ranking scheme to present only the methodology; however, other corrections for all-vs-all pairwise comparisons, such as *Shaffer's correction*, can be used [48].

6.1.2 Examples

6.1.2.1 Data

To show how the DSC ranking scheme works and to compare it with already-existing ranking schemes for benchmarking single-objective stochastic optimization algorithms, the algorithms presented at the sixth Black-Box Optimization Benchmarking 2015 (i.e., BBOB 2015) workshop [9] are used in the comparison.

BBOB 2015 was a competition for benchmarking algorithms on single-objective problems [52]. The test problems are organized into five groups:

- separable functions,
- functions with low or moderate conditioning,
- functions with high conditioning and unimodal,
- multi-modal functions with an adequate global structure,
- multi-modal functions with a weak global structure.

Fifteen algorithms were selected for the examples: BSif [88], BSifeg [88], BSqi [88], BSrr [88], CMA-CSA [3], CMA-MSR [3], CMA-TPA [3], GP1-CMAES [4], GP5-CMAES [4], RAND-2xDefault [13], RF1-CMAES [4], RF5-CMAES [4], Sif [88], Sifeg [88], and Srr [88]. The selected algorithms were part of the BBOB 2015 competition. They were selected for illustrative purposes only.

For the presented examples, all the comparisons are on 22 different noiseless problems with the dimension fixed at 10. We must point out here that in the presented examples the analysis is performed at a problem level, and we do not distinguish between different problem instances from the same problem. Two examples will be presented. In the first example, a comparison between three randomly selected algorithms is presented, while in the second example, a scenario of multiple comparisons with a control algorithm is presented, where all 15 algorithms are involved in the comparison.

Both examples are performed with the common and DSC approaches. The common approach is used twice concerning the representative value selected to represent the algorithm's multiple runs on a single problem (i.e., the mean and median). The DSC ranking scheme is also used twice concerning the statistical test that is used for comparing one-dimensional distributions (i.e., the two-sample KS and the two-sample AD test).

6.1.2.2 Multiple Comparisons Among Algorithms

Here, examples that involve comparisons of three randomly selected algorithms are presented. Table 6.1 presents the p-values of the comparisons between three combinations of three algorithms. After obtaining the rankings, no matter which ranking scheme is used, the appropriate statistical omnibus test is the Friedman test.

Using the results presented in the table, it follows that the results from the first combination using the common approach differ from the results obtained with the DSC approach. The common approach rejects the null hypothesis, so there is a statistical significance between the performances of the compared algorithms. However, the opposite is true when the DSC rankings scheme is used i.e., there is no statistical significance between the performance of the algorithms.

Looking into the p-values obtained with the common approach or the DSC approach, it is obvious that the results are not statistically different (i.e., from the perspective of the decision about the null hypothesis); they obtained different p-values when different criteria are involved in the ranking schemes. These results indicate that different statistics, which are used to find the representative value from the multiple runs of each algorithm on each problem, affect the calculation of the test statistic that is involved with the omnibus statistical test, and they lead to different p-values [24].

Comparing the statistical outcome in the second and third combinations, it follows that the result is statistically the same. For the second combination, the null hypothesis is not rejected using both approaches with different statistical criteria. While in the third combination, the null hypothesis is rejected for both approaches.

To show why the differences can be presented when different approaches are used, all the steps involved in the comparisons performed for the first combination are presented in more detail. Table 6.2 presents the rankings obtained for each compared algorithm on each problem involved in the benchmarking using both approaches (i.e., the common and the DSC) for different statistical criteria. Using the results presented in the table, it is obvious that there are problems for which the algorithms obtained

Table 6.1 Statistical comparisons of three algorithms

Algorithms	p_{value}			
	Common approach (with means)	Common approach (with medians)	DSC approach (KS)	DSC approach (AD)
GP5-CMAES, Sifeg, BSif	*(0.02)	*(0.04)	(0.42)	(0.44)
BSifeg, RF1-CMAES, BSrr	(0.16)	(0.23)	(0.28)	(0.48)
BSrr, RAND-2xDefault, Srr	*(0.00)	*(0.00)	*(0.00)	*(0.00)

* Indicates that the null hypothesis is rejected, using $\alpha = 0.05$
p_{value} corresponds to the p-value obtained by the Friedman test

Table 6.2 Rankings for the algorithms A_1 = GP5-CMAES, A_2 = Sifeg, and A_3 = BSif

F	Common approach (with means)			Common approach (with medians)			DSC approach (KS)			DSC approach (AD)		
	A_1	A_2	A_3	A_1	A_2	A_3	A_1	A_2	A_3	A_1	A_2	A_3
f_1	3.00	2.00	1.00	3.00	2.00	1.00	3.00	2.00	1.00	3.00	2.00	1.00
f_2	3.00	2.00	1.00	3.00	2.00	1.00	3.00	2.00	1.00	3.00	2.00	1.00
f_3	3.00	2.00	1.00	3.00	2.00	1.00	3.00	2.00	1.00	3.00	2.00	1.00
f_4	3.00	1.00	2.00	3.00	1.00	2.00	3.00	1.00	2.00	3.00	1.00	2.00
f_5	3.00	1.50	1.50	2.00	2.00	2.00	2.00	2.00	2.00	2.00	2.00	2.00
f_6	3.00	1.00	2.00	3.00	1.00	2.00	3.00	1.00	2.00	2.50	1.00	2.50
f_7	1.00	2.00	3.00	1.00	2.00	3.00	1.00	2.50	2.50	1.00	2.50	2.50
f_8	3.00	1.00	2.00	3.00	1.00	2.00	3.00	1.50	1.50	3.00	1.00	2.00
f_9	3.00	1.00	2.00	3.00	1.00	2.00	3.00	1.50	1.50	3.00	1.50	1.50
f_{10}	1.00	2.00	3.00	1.00	2.00	3.00	1.00	2.50	2.50	1.00	2.50	2.50
f_{11}	1.00	2.00	3.00	1.00	2.00	3.00	1.00	2.50	2.50	1.00	2.50	2.50
f_{12}	3.00	2.00	1.00	3.00	2.00	1.00	3.00	1.50	1.50	3.00	1.50	1.50
f_{13}	**2.00**	**1.00**	**3.00**	**1.00**	**2.00**	**3.00**	**1.50**	**1.50**	**3.00**	**1.50**	**1.50**	**3.00**
f_{14}	3.00	1.00	2.00	1.00	2.00	3.00	2.50	2.50	1.00	2.50	2.50	1.00
f_{15}	2.00	1.00	3.00	2.00	1.00	3.00	2.00	2.00	2.00	2.00	2.00	2.00
f_{16}	2.00	1.00	3.00	2.00	1.00	3.00	2.00	2.00	2.00	2.00	1.00	3.00
f_{17}	1.00	2.00	3.00	1.00	2.00	3.00	1.00	2.50	2.50	1.00	2.50	2.50
f_{18}	**1.00**	**2.00**	**3.00**	**1.00**	**2.00**	**3.00**	**1.00**	**2.50**	**2.50**	**1.00**	**2.00**	**3.00**
f_{19}	3.00	1.00	2.00	3.00	1.00	2.00	3.00	1.50	1.50	3.00	1.50	1.50
f_{20}	3.00	1.00	2.00	3.00	1.00	2.00	3.00	1.50	1.50	3.00	1.50	1.50
f_{21}	1.00	2.00	3.00	1.00	2.00	3.00	2.00	2.00	2.00	2.00	2.00	2.00
f_{22}	**1.00**	**2.00**	**3.00**	**2.00**	**1.00**	**3.00**	**2.00**	**2.00**	**2.00**	**2.00**	**2.00**	**2.00**

different rankings when a different approach is used (i.e., the common or the DSC). Furthermore, it can happen that the rankings differ when the same approach is used with different statistical criteria. To go into more detail, let us present some examples at the single-problem level.

Looking into the 13th problem, f_{13}, the obtained rankings using the common approach are 2.00, 1.00, 3.00, when means are involved to select the representative values, and 1.00, 2.00, 3.00, when medians are involved. From the results it follows that the common approach with different statistical criteria (i.e., the mean or the median) leads to different conclusions. In the case of the DSC approach, the obtained rankings are 1.50, 1.50, 3.00 using both criteria for comparing the distributions of the obtained solution values (i.e., the two-sample KS test and the two-sample AD test). Furthermore, the DSC rankings differ from the rankings obtained by the common approach irrespective of which statistics is used.

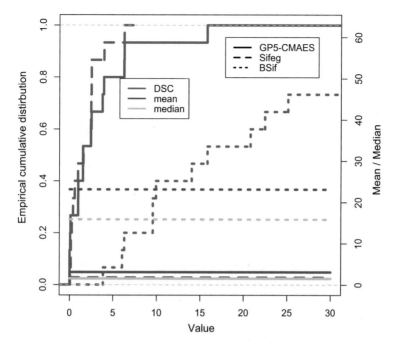

Fig. 6.1 Empirical cumulative distributions (step functions) and mean and median values (horizontal lines) for f_{13}

Table 6.3 Descriptive statistics obtained on f_{13}

Algorithm	Mean	Median
GP5-CMAES	3.07	1.53
Sifeg	1.77	1.57
BSif	23.11	15.84

To explain what is happening in the ranking processes and why these differences appear, Fig. 6.1 presents the mean and the median values of the solution values from multiple runs of each algorithm obtained on f_{13} (purple and green lines, respectively), and the empirical cumulative distributions of the obtained solution values from the multiple runs of each algorithm obtained on f_{13} (red (step) lines). Using the common approach, a smaller value is better since a minimization problem is being considered. From Fig. 6.1, it is clear that the rankings with regard to the mean values are 2.00, 1.00, and 3.00. However, using the median values, we cannot draw a clear conclusion, since the median values for the first and second algorithms (i.e., GP5-CMAES and Sifeg) are overlapping. For this purpose, Table 6.3 presents the mean and median values of the solution values from multiple runs for each algorithm on f_{13}.

Using the results presented in the table, it follows that GP5-CMAES has a mean of 3.07 and Sifeg has a mean of 1.77, while the medians are 1.53 and 1.57, respectively.

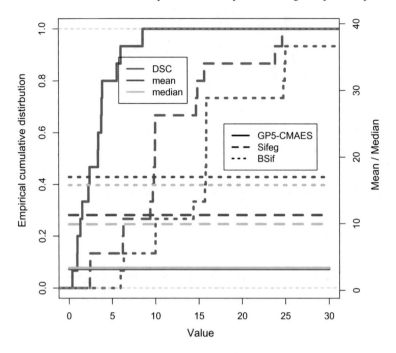

Fig. 6.2 Empirical cumulative distributions (step functions) and the mean and median values (horizontal lines) for f_{18}

Since a smaller value is better, they swap their rankings using different statistical criteria in the same approach. This happens because the mean can be affected by outliers (i.e., poor runs that exist) and this actually happens for the GP5-CMAES algorithm. Using the empirical cumulative distribution, it follows that there are some poor runs around the value of 15 that affect the mean of the obtained solution of the multiple runs obtained using the GP5-CMAES. In the case when medians are used (i.e., more robust statistics), it follows that the medians for the GP5-CMAES and Sifeg are in a small ϵ-neighborhood; however, they obtain different rankings with the fractional ranking scheme, since they are different. These problems can be omitted with the DSC ranking scheme because it works by comparing whole distributions and not only one statistic that describes the distribution.

For the 18th problem, f_{18}, Fig. 6.2 presents the means, medians, and empirical cumulative distributions of the solution values from multiple runs of the three algorithms.

From the figure, it follows that the common approach with either means or medians provides the same rankings, while the DSC approach with different criteria for comparing one-dimensional distributions provides different rankings.

The DSC ranking scheme involves all the pairwise comparisons between the compared algorithms. When using the two-sample KS test the obtained p-values are 0.00 (GP5-CMAES, Sifeg), 0.00 (GP5-CMAES, BSif), and 0.03 (Sifeg, BSif). Further-

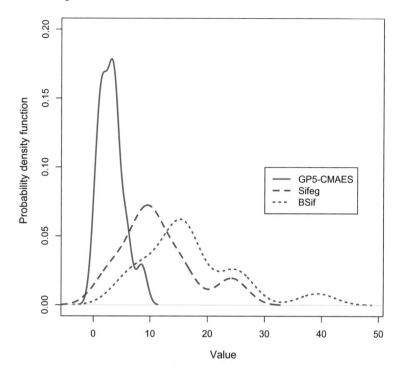

Fig. 6.3 Empirical probability density functions for multiple runs of each algorithm obtained on f_{18}

more, these p-values are corrected using the Bonferroni correction, the transitivity of the p-values matrix is satisfied and the algorithms are split into two disjointed sets {GP5-CMAES} and {Sifeg, BSif}. Next, concerning the information about which set the algorithms belong to, their rankings are assigned as 1.00, 2.50, and 2.50.

When using the two-sample AD test, the p-values for the pairwise comparisons are 0.00 (GP5-CMAES, Sifeg), 0.00 (GP5-CMAES, BSif), and 0.01 (Sifeg, BSif). When correcting them with the Bonferroni correction, it follows that the transitivity of the p-values matrix is not satisfied, so the algorithms obtained their rankings according to their averages of the obtained solutions values (i.e., lower value means better ranking).

The question that arises here is why the two different criteria for comparing the distributions provide different results [33]. For this purpose, Fig. 6.3 presents the probability density functions of the solution values from multiple runs for each algorithm obtained on f_{18}. The figure indicates that the two-sample KS test is not able to detect the difference that exists in the shift and tail ends between the distributions of Sifeg and BSif. These results indicate that it is better to use the two-sample AD test as a criterion in the DSC ranking scheme, since it is more powerful and it can better detect differences than the *two-sample KS test* when distributions vary in shift

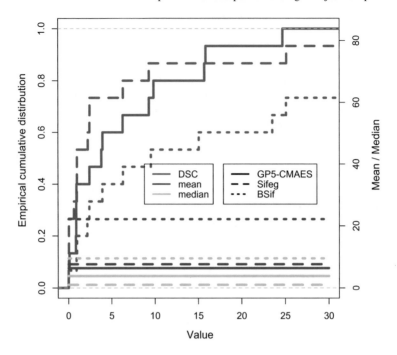

Fig. 6.4 Empirical cumulative distributions (step functions) and mean and median values (horizontal lines) for f_{22}

Table 6.4 Descriptive statistics obtained on f_{22}

Algorithm	Mean	Median
GP5-CMAES	6.37	3.82
Sifeg	7.63	0.99
BSif	22.22	9.52

only, in symmetry only, or have the same mean and standard deviation, but differ in the tail ends only. This result is known and supported in statistical literature [42].

Focusing on the 22nd problem, f_{22}, Fig. 6.4 presents the means, medians, and empirical cumulative distributions of the obtained solution values from the multiple runs of the algorithms. Table 6.4 presents the means and the medians of the obtained solutions from the multiple runs for each algorithm obtained on f_{22}, which are used by the common approach. Using the DSC ranking scheme, both statistical criteria used for comparing distributions give the same rankings, i.e., $2.00, 2.00$, and 2.00. From the figure, it is not clear whether the presented distributions are statistically different. For this purpose, Table 6.5 presents the p-values obtained for all the pairwise comparisons using both statistical tests for comparing the distributions. Furthermore, these p-values are corrected using the Bonferroni correction, the transitivity is satisfied and all the algorithms belong to one set, so they obtain the same ranking.

Table 6.5 P-values for all pairwise comparisons involved in the DSC ranking scheme obtained on f_{22}

Pairs of algorithms	p_{value}	
	KS	AD
(GP5-CMAES, Sifeg)	0.67	0.51
(GP5-CMAES, BSif)	0.38	0.09
(Sifeg, BSif)	0.18	0.06

We must point out that if the benchmarking result detects a difference between the performance of the compared algorithms (i.e., the null hypothesis is rejected), post-hoc statistical test should be performed to see from where this difference comes. However, we omitted this step because we would like to point out only the impact of the statistics on the benchmarking in metaheuristic stochastic optimization algorithms. The influence that appears in the omnibus statistical test is also transferred at the post-hoc level.

6.1.2.3 Multiple Comparisons with a Control Algorithm

Here, an example of multiple comparisons with a control algorithm is presented, which is the most commonly used scenario in publishing scientific articles. This involves comparing the performance of a newly developed algorithm with the performances of state-of-the-art algorithms. To use the DSC ranking scheme in this scenario, it is investigated using multiple pairwise comparisons performed using the *Wilcoxon test*. This is done because it can happen that the DSC ranking scheme is affected by the p-value correction when the number of compared algorithms increases.

Table 6.6 presents the p-values for each pairwise comparison using both approaches (i.e., the common and the DSC) with different statistical criteria. Using the p-values presented, it is obvious that different statistical criteria lead to different test statistic values. Focusing on one pairwise comparison performed between CMA-CSA and CMA-MSR, the DSC ranking scheme with both statistical criteria does not reject the null hypothesis, the common approach with means rejects the null hypothesis, while the common approach with medians also shows that the null hypothesis is not rejected.

Let us focus on the results obtained with the DSC ranking scheme using the two-sample KS test. It follows that the performance of the CMA-CSA is statistically significant compared to the performance of BSif, BSifeg, BSqi, BSrr, GP1-CMAES, GP5-CMAES, RAND-2xDefault, RF1-CMAES, RF5-CMAES, Sif, Sifeg, and Srr. It also happens that its performance is not statistically significant with regard to the performances of CMA-MSR and CMA-TPA. However, we must point out here that these statistical results are only valid assuming independent pairwise comparisons.

Table 6.6 Multiple comparisons with a control algorithm (CMA-CSA) by using multiple Wilcoxon tests

CMA-CSA versus	p_{value}			
	(DSC; KS)	(DSC; AD)	(CA; mean)	(CA; median)
BSif	4.847534e-03	4.847534e-03	8.476892e-04	8.476892e-04
BSifeg	7.768118e-03	7.768118e-03	1.086096e-03	1.758873e-03
BSqi	3.081757e-03	7.768118e-03	1.227287e-03	2.223195e-03
BSrr	7.768118e-03	7.768118e-03	1.086096e-03	1.758873e-03
CMA-MSR	1.000000e+00	7.655945e-01	4.757041e-02	7.628835e-02
CMA-TPA	1.000000e+00	3.457786e-01	4.757041e-02	4.654448e-01
GP1-CMAES	1.451271e-05	8.553503e-06	4.768372e-07	6.411516e-05
GP5-CMAES	8.553503e-06	8.553503e-06	4.768372e-07	6.411516e-05
RAND-2xDefault	5.049088e-06	5.049088e-06	6.411516e-05	6.411516e-05
RF1-CMAES	5.049088e-06	5.049088e-06	4.768372e-07	6.411516e-05
RF5-CMAES	5.049088e-06	5.049088e-06	4.768372e-07	6.411516e-05
Sif	6.301490e-04	3.759531e-04	9.600603e-04	1.385265e-03
Sifeg	6.301490e-04	6.301490e-04	1.385265e-03	3.504330e-03
Srr	1.056542e-03	6.301490e-04	1.227287e-03	2.495261e-03

In the cases when we are interested in comparing one vs. all (i.e., to check if there is a statistical significance between the performance of CMA-CSA and the other 12 algorithms), the true p-value must be calculated by combining the p-values of the independent comparisons using the following equation

$$p_{value} = 1 - \prod_{i=1}^{k-1}[1 - p_{value_i}], \qquad (6.1)$$

where $k - 1$ is the number of independent pairwise comparisons that are combined. In our case $k - 1 = 12$, since we are comparing our control algorithm with 12 other algorithms for which a statistical significance is presented in the independent pairwise comparisons. The number of algorithms in this example is $k = 13$, since both algorithms for which the pairwise independent comparisons are not statistically significant are omitted from the comparison.

Using this equation it follows that the p-value is 0.02, so the CMA-CSA has a statistically significant performance compared to the algorithms: BSif, BSifeg, BSqi, BSrr, GP1-CMAES, GP5-CMAES, RAND-2xDefault, RF1-CMAES, RF5-CMAES, Sif, Sifeg, and Srr.

The same analysis can be performed using the results obtained using the common approach with means and medians, and by the DSC approach with the two-sample AD test.

6.2 Practical Significance

The statistical significance is crucial when developing a new algorithm in academia, since the newly developed algorithm should have a statistically significant, improved performance compared to state-of-the-art algorithms. However, there is still a large gap between academia and real-world scenarios. This happens because some differences that indicate the presence of a statistical significance are not significant in a practical sense [66]. The practical significance provides information about the quality of solutions in real-world applications [65, 82].

It is, therefore, important to understand the difference between practical significance and statistical significance [28]. For instance, let's compare two algorithms designed to find the optimum solution to a given problem. The first algorithm solves the problem with an approximation error of 10^{-10}, while the second one has an error of 10^{-16}. Although a statistical significance can be found when comparing the outcomes between these two algorithms, this difference can be insignificant in a practical sense with respect to the application of the problem.

To define the practical significance, let us assume that two data values are not different for some real-world application scenario if the difference between them is smaller than or equal to some predefined threshold ϵ. Furthermore, we can mathematically define the practical difference as $f_\epsilon(x_1, x_2)$, where x_1 and x_2 are two data values (i.e., in our case estimations of the quality of the solutions):

$$f_\epsilon(x_1, x_2) = \begin{cases} x_1 = x_2, & \text{if } x_1 < x_2 + \epsilon \text{ and } x_1 > x_2 - \epsilon \\ x_1 \neq x_2, & \text{otherwise} \end{cases}. \tag{6.2}$$

For a given real-world problem, the predefined practical threshold should be defined by a domain expert of the problem being solved. From here it follows that the predefined practical threshold, ϵ, can be different for different optimization problems.

There are a lot of use cases in industry and academia where practical significance should be addressed. First, let us mention some examples where practical significance should be checked in different industrial tasks. In production, many items are created with some tolerance deviations that do not affect the performance of the end product, which means that all the products created within these deviations are assumed to be equal. An example of this can be found in [70], where the parametrization of an electric motor's design (i.e., its rotor and stator) is investigated to find its geometrical characteristics. A lot of industrial tasks also involve simulation tools, where the simulation results differ from the real-world performance. These differences between the results at the level of some decimal points might not affect the design of the end product. Another example is the selection of an algorithm that should be used for a specific problem. Let us assume that we can select between two algorithms, where the first algorithm has a more rapid convergence, but the second algorithm provides better results. In the case when the results are not practically significant, we should select the first algorithm because it is faster.

Second, let us mention some examples where practical significance should be considered in academia. The benchmark experiments where a comparison of algorithms' performances is made should address this issue with great care. Nowadays, there are several competitions such as Black-Box Optimization Benchmarking (BBOB) [53], where the results can be affected by the calculation process of the test problems (i.e., the influence of the computer's accuracy due to the IEEE 754 standard (e.g., values closer to zero can be more accurately presented), the variables type (e.g., 4-byte float, 8-byte float, 10-byte float), or the stopping criteria threshold for different algorithms. So, taking care of these issues, we can estimate the actual performance, which will not be the case if only the statistical significance is identified.

6.2.1 Practical Deep Statistical Comparison Ranking Scheme

To find out if some optimization results are practically significant or not, we modified the DSC ranking scheme. The main idea behind the *practical Deep Statistical Comparison (pDSC)* approach is that the search for practical significance is made as a pre-processing step before analyzing the distribution of the results, as is the case with the DSC ranking scheme. For benchmarking purposes, two pDSC ranking schemes were proposed concerning how the practical significance check is performed. The first one is known as a *sequential pDSC ranking scheme*, where the pre-processing is in a sequential order, where the gth run from one algorithm is compared with the gth run of the other algorithm, $g = 1, \ldots, n$, considering a predefined practical threshold, ϵ, and the Eq. 6.2. One weakness of such pre-processing is the sequential order of the calculations, since these algorithms are stochastic in nature and there is no guarantee that the same order of values will be produced if the algorithms are run again. This can affect the result of practical significance testing. To go beyond this, the second pDSC ranking scheme has been proposed, which is known as *Monte Carlo pDSC*. This ranking scheme performs a Monte Carlo simulation for each pairwise comparison involved in the DSC ranking scheme, where permutations of the independent runs of both algorithms are involved in the comparison. It simulates n runs of the algorithms where the final solutions are obtained in a different order. The same as with the sequential pDSC, the practical threshold, ϵ should be provided a priori by a domain expert. When the practical threshold is set to 0, both pDSC ranking schemes become the DSC ranking scheme.

6.2.1.1 Sequential Practical Deep Statistical Comparison

In the sequential version of the pDSC, the pre-processing is in the same sequence as the independent runs from the algorithms are obtained. As a first step, the practical threshold, ϵ_p, should be set by a domain expert or user. The pre-processing step comes before comparing the distributions of the results between each pair of algorithms involved in the comparison. This means that first we should check the data for the practical threshold and then the matrix defined in Eq. 5.1 for the DSC ranking scheme should be calculated. To illustrate how the pre-processing is performed, let us assume that n independent results are obtained for two algorithms A and B on the same optimization problem. So before comparing the distributions of the obtained results, the data is pre-processed as follows:

$$
\begin{cases}
a_g = b_g = \frac{a_g + b_g}{2}, & |a_g - b_g| \le \epsilon_p \\
a_g = a_g, b_g = b_g, & |a_g - b_g| > \epsilon_p
\end{cases}
\tag{6.3}
$$

where $g = 1, \ldots, n$, represents one independent run, and a_g and b_g are the obtained solution values from the A and B algorithm in the gth run, respectively. From this equation, it follows that two data values that are the results from the two algorithms for the same run should be swapped with their mean value if their absolute difference is smaller than or equal to the predefined threshold, ϵ_p, while otherwise they will stay the same without any transformation. This kind of pre-processing should be applied for each pair of algorithms that are involved in the comparison and for each optimization problem separately.

The pre-processed data for each pairwise comparison is further used in Eq. 5.1 of the DSC ranking scheme to define the p-value for each pairwise comparison. The following steps remain the same as in the DSC ranking scheme: the p-values are corrected to control the FWER, the transitivity of the matrix is checked, and the rankings of the algorithms are assigned.

The pipeline of the sequential pDSC ranking scheme is presented in Fig. 6.5. The difference between the data values obtained for the same runs that are smaller than or equal to the predefined threshold is marked with *. To run this scheme, all the algorithms should have the same number of runs. In cases where the number of runs differ, then some sampling techniques should be used to ensure that all the algorithms have the same number of runs. For each algorithm, the missing values should be sampled from the distribution of the obtained results.

6.2.1.2 Monte Carlo Practical Deep Statistical Comparison

The sequential pDSC ranking scheme depends on and can be affected by the order of the independent runs that are obtained using the algorithms. To go beyond this, the Monte Carlo pDSC ranking scheme has been proposed. So, before each pairwise comparison is performed, for each algorithm different orders of the obtained results are generated using its permutations. The number of such permutations is $n!$. Furthermore, for each pairwise comparison that includes the algorithms A and B, $(n!)^2$ different combinations exist, which can be used to check the practical significance. Each combination has one permutation of the set of multiple runs from the first algorithm and one permutation from the set of multiple runs from the second algorithm. To find the p-value associated with each pairwise comparison required for Eq. 5.1, for each one N different combinations out of $(n!)^2$ are randomly selected. Next, for each

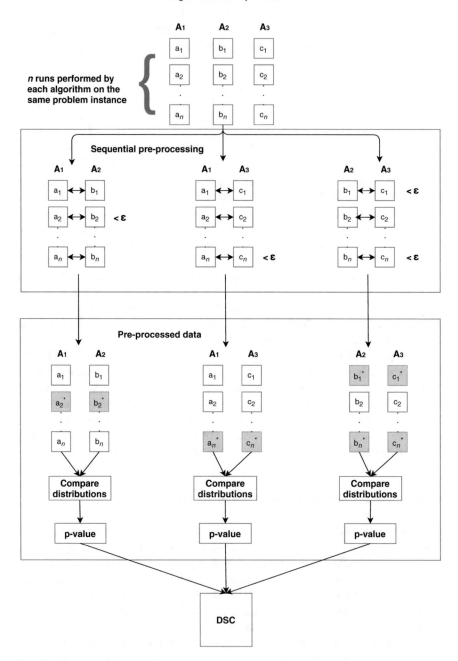

Fig. 6.5 Flowchart for sequential pDSC [25]

selected combination, the pre-processing uses Eq. 6.3 and after that the distributions are compared using a selected test for comparing the whole distributions. The results of this process are N p-values from which one should be selected to be involved in the Eq. 5.1. To find an appropriate p-value from the N p-values, the mode of the p-values distribution can be used. However, it can happen that the distribution of the p-values is multi-modal, so a question arises as to which one is selected. For this purpose, a new random variable V is introduced, which is the number of combinations where the null hypothesis is rejected. Furthermore, a probability threshold α_p should be set that gives us the information when the distributions are considered as different concerning the definition of V (see Eq. 6.4).

$$
\begin{cases}
P(V) < \alpha_p, & A \text{ and } B \text{ have the same distribution} \\
P(V) \geq \alpha_p, & A \text{ and } B \text{ have different distributions}
\end{cases}
\tag{6.4}
$$

Using Eq. 6.4, we can detect whether the compared distributions are the same or different. If the distributions are the same, then the p-value for this pairwise comparison can be randomly selected from a subset of N p-values that are larger than the significance level used for the statistical test for comparing distributions, α. If the distributions are different, a kernel density estimation [94] should be used to estimate the probability density function of a subset of N p-values that are smaller than α. Next, the mode of the probability density function is used as an appropriate p-value, which will be used in Eq. 5.1. This voids the selection of a random p-value that will be further affected in the p-values correction used to control the FWER. The pipeline of the Monte Carlo pDSC ranking scheme is presented in Fig. 6.6.

6.2.2 Examples

Table 6.7 shows the results obtained when comparing one combination of three algorithms using the sequential variant of the pDSC with different practical significance levels. The data is the same as used to present how the DSC ranking scheme works (see Sect. 5.2.1). For the simplicity of our experiments, we set the same practical threshold for all 22 optimization problems that are involved in the comparison. However, users can define different practical thresholds for different optimization problems. The results presented in Table 6.7 indicate that when the ϵ threshold exceeds a certain value, the outcome of the hypothesis testing changes from "different" (denoted by 0) to "equal" (denoted by 1). Different means that there is a practical significance between the performance of the compared algorithms using a set of 22 benchmark problems, while equal means the opposite. This result actually shows that the practical significance of the compared algorithms depends on the practical threshold involved in the comparison.

To show why the differences appear, let us focus in more detail on an example where different practical thresholds are involved. For this purpose, Table 6.8 presents

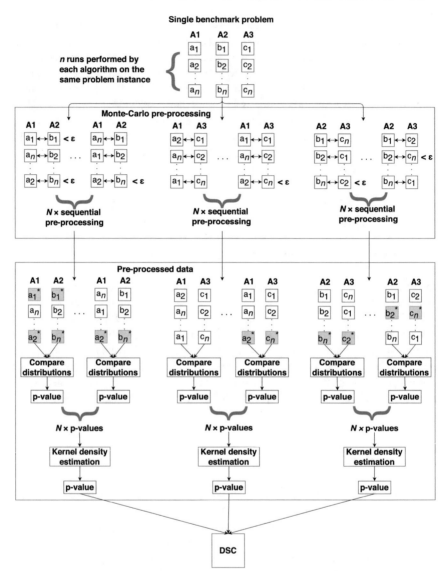

Fig. 6.6 Flowchart for Monte Carlo pDSC [25]

Table 6.7 Statistical comparison of three algorithms using the sequential variant of the pDSC

Algorithms	ϵ_p							
	10^{-9}	10^{-6}	10^{-3}	10^{-2}	10^{-1}	10^0	10^1	10^2
RF1-CMAES, GP1-CMAES, Srr	0	0	0	1	1	1	1	1

0-Indicates that the null hypothesis is rejected, $p_{value} < 0.05$
1-Indicates that the null hypothesis fails to reject, $p_{value} \geq 0.05$
p_{value} corresponds to the p-value obtained by the Friedman test

Table 6.8 Rankings for the algorithms A_1 = RF1-CMAES, A_2 = GP1-CMAES, and A_3 = Srr, using the sequential variant of the pDSC ranking scheme

F	$\epsilon_p = 10^{-1}$			$\epsilon_p = 10^0$			$\epsilon_p = 10^1$		
	A_1	A_2	A_3	A_1	A_2	A_3	A_1	A_2	A_3
f_1	2.00	2.00	2.00	2.00	2.00	2.00	2.00	2.00	2.00
f_2	3.00	2.00	1.00	3.00	2.00	1.00	3.00	2.00	1.00
f_3	2.50	2.50	1.00	2.50	2.50	1.00	2.50	2.50	1.00
f_4	3.00	2.00	1.00	3.00	2.00	1.00	3.00	2.00	1.00
f_5	2.00	2.00	2.00	2.00	2.00	2.00	2.00	2.00	2.00
f_6	3.00	1.50	1.50	3.00	1.50	1.50	3.00	1.50	1.50
f_7	2.50	1.00	2.50	2.50	1.00	2.50	2.00	2.00	2.00
f_8	3.00	2.00	1.00	3.00	2.00	1.00	2.00	2.00	2.00
f_9	3.00	1.50	1.50	3.00	1.50	1.50	3.00	1.50	1.50
f_{10}	2.50	1.00	2.50	2.50	1.00	2.50	2.50	1.00	2.50
f_{11}	2.50	1.00	2.50	2.50	1.00	2.50	2.50	1.00	2.50
f_{12}	2.00	2.00	2.00	2.00	2.00	2.00	2.00	2.00	2.00
f_{13}	3.00	1.50	1.50	3.00	1.50	1.50	3.00	2.00	1.00
f_{14}	2.00	2.00	2.00	2.00	2.00	2.00	2.00	2.00	2.00
f_{15}	1.50	1.50	3.00	1.50	1.50	3.00	1.50	1.50	3.00
f_{16}	3.00	2.00	1.00	3.00	2.00	1.00	2.00	2.00	2.00
f_{17}	2.00	1.00	3.00	2.00	1.00	3.00	2.00	2.00	2.00
f_{18}	**2.00**	**1.00**	**3.00**	**1.50**	**1.50**	**3.00**	**2.00**	**2.00**	**2.00**
f_{19}	2.00	3.00	1.00	2.00	3.00	1.00	2.00	2.00	2.00
f_{20}	2.50	2.50	1.00	2.50	2.50	1.00	2.00	2.00	2.00
f_{21}	2.00	2.00	2.00	2.00	2.00	2.00	2.00	2.00	2.00
f_{22}	2.00	2.00	2.00	2.00	2.00	2.00	2.00	2.00	2.00

the rankings obtained when comparing the algorithms RF1-CMAES, GP1-CMAES, and Srr for different practical thresholds $\epsilon_p \in \{10^{-1}, 10^0, 10^1\}$ using the sequential version of the pDSC ranking scheme. The rankings obtained for $\epsilon_p = 10^{-1}$ and $\epsilon_p = 10^0$ differ only in one problem, f_{18}. While when the practical threshold increases to $\epsilon_p = 10^1$, the rankings differ in eight problems from the rankings obtained when $\epsilon_p = 10^{-1}$ and $\epsilon_p = 10^0$.

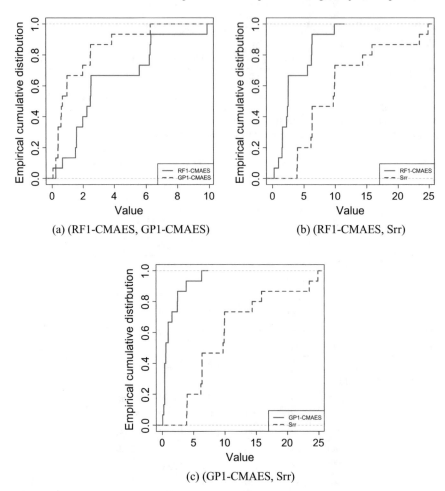

(a) (RF1-CMAES, GP1-CMAES)   (b) (RF1-CMAES, Srr)

(c) (GP1-CMAES, Srr)

Fig. 6.7 Empirical cumulative distributions for f_{18} at a practical significance level of 10^0

Let us focus on a single-problem level. For this purpose, the problem f_{18} is selected. The difference in the ranking happens because the comparison is made by considering the whole distributions of the obtained solution values and how different pre-processing thresholds change the data distributions. Figures 6.7 and 6.8 present the empirical cumulative distributions for all the pairwise comparisons between the three algorithms RF1-CMAES, GP1-CMAES, and Srr, for two different practical significance levels. The x-axis, "Value", represents the achieved error of the optimization runs.

When the practical threshold is set to $\epsilon_p = 10^0$ (Table 6.8), the rankings are 1.50, 1.50, 3.00, respectively. The p-values for the pairwise comparisons are 0.021 for (RF1-CMAES, GP1-CMAES), 0.000 for (RF1-CMAES, Srr), and 0.000 for (GP1-CMAES, Srr), which are further tested at a significance level of 0.016, considering the

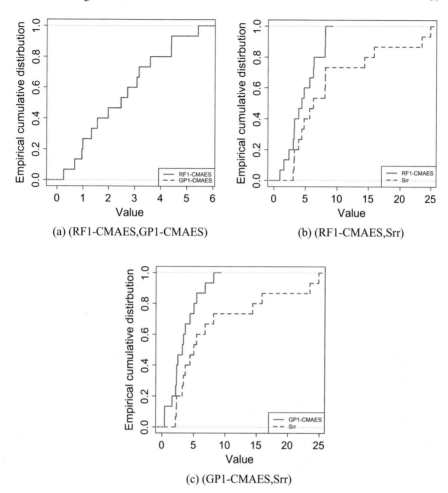

Fig. 6.8 Empirical cumulative distributions for f_{18} at a practical significance level of 10^1

Bonferroni correction. The transitivity of the p-values matrix is satisfied, so they are split into two disjoint sets {RF1-CMAES, GP1-CMAES} and {Srr}. Furthermore, the algorithms obtain a ranking from the set to which they belong.

When the practical threshold increases to $\epsilon_p = 10^1$ (Table 6.8), the rankings are 2.00, 2.00, and 2.00, respectively. The p-values obtained for the pairwise comparisons are 1.000 for (RF1-CMAES, GP1-CMAES), 0.08 for (RF1-CMAES, Srr), and 0.06 for (GP1-CMAES, Srr). From the obtained p-values and using Fig. 6.8, it is obvious that the distributions are the same. From here, it follows that all the algorithms were given the same ranking.

To see what happens when the order of the independent runs changes, a different order permutation for problem f_{18} is used for each algorithm (see Table 6.9). Here, the practical threshold is set to $\epsilon_p = 10^0$. Using the permuted data, the rankings

Table 6.9 Permutations of independent runs obtained for problem f_{18} by the algorithms $A_1 =$ RF1-CMAES, $A_2 =$ GP1-CMAES, and $A_3 =$ Srr

A_1	A_2	A_3
9.864	2.439	9.862
2.509	0.971	9.693
6.259	0.387	6.309
1.579	3.848	3.970
6.208	0.396	9.908
1.510	0.579	6.307
0.250	0.397	24.844
6.308	0.966	15.842
2.381	0.056	6.305
5.576	1.577	3.882
0.957	0.604	23.460
2.244	0.391	6.121
2.490	6.271	14.341
1.584	0.244	3.930
2.497	2.483	9.937

Table 6.10 Statistical comparison of three algorithms using the Monte Carlo variant of the pDSC

Algorithms	ϵ_p							
	10^{-9}	10^{-6}	10^{-3}	10^{-2}	10^{-1}	10^{0}	10^{1}	10^{2}
RF1-CMAES, GP1-CMAES, Srr	0	0	0	0	0	0	1	1

0-Indicates that the null hypothesis is rejected, $p_{value} < 0.05$
1-Indicates that the null hypothesis fails to reject, $p_{value} \geq 0.05$
p_{value} corresponds to the p-value obtained by the Friedman test

obtained using the sequential variant of the pDSC ranking scheme are 2.00, 1.00, and 3.00, respectively. With the original order of the obtained solution values for the same practical threshold, the obtained rankings were 1.50, 1.50, and 3.00. From here, it follows that the order of independent runs influences the pre-processing, which affects the practical significance.

To avoid the dependence on the order of the independent runs, the same combination of the three algorithms was analyzed using the set of 22 benchmark problems by applying the Monte Carlo pDSC ranking scheme for the same range of the practical thresholds presented in Table 6.7. Table 6.10 presents the p-values obtained using the Monte Carlo pDSC ranking scheme. Comparing them with the p-values obtained using the sequential pDSC ranking scheme, there are differences that happen for some parasitical thresholds: 10^{-2}, 10^{-1}, and 10^{0}. This result indicates that for these practical thresholds the order of the independent runs affects the ranking process.

In case of the Monte Carlo pDSC, for each algorithm, the number of independent runs provided by the BBOB 2015 for each problem is 15, $n = 15$. Furthermore, 15! permutations can be generated for each algorithm on a single problem. For each pairwise comparison $(15!)^2$ different combinations exist that can be involved in the comparison. However, for each one we randomly selected 1000 out of $(15!)^2$. Further, the two-sample AD test is used to obtain a p-value for each of the 1000 selected combinations. To select one p-value from the 1000 p-values that will be assigned to the pairwise comparison, Eq. 6.4 is used to select a p-value with a prior level of significance, $\alpha_p = 0.05$.

To show the applicability of the Monte Carlo pDSC on a single problem, the problem f_{18} is investigated in more detail, when $\epsilon_p = 10^0$. For each pairwise comparison, we selected 1000 permutations of the 15 independent runs for each algorithm on the problem f_{18}. Each combination consists of one permutation from the first algorithm and one permutation from the second algorithm. Further, the data is pre-processed using Eq. 6.3. This data is then used by the two-sample AD test. After processing 1000 different combinations, the ranking scheme uses Eq. 6.4 to select one p-value according to a prior level of significance, α_p.

Figure 6.9 presents the kernel density estimation for the p-values of each pairwise comparison. The blue vertical line corresponds to the mode of the probability density function that is used from Eq. 6.4. The selected p-value is 0.008 for (RF1-CMAES, GP1-CMAES), 0.000 for (RF1-CMAES, Srr), and 0.000 for (GP1-CMAES, Srr). The Bonferroni correction is then used to correct the p-values, the transitivity of the p-values matrix is not satisfied and the obtained rankings are 2.00, 1.00, and 3.00, respectively. This result is different from the one obtained by the sequential variant of the pDSC.

6.3 Extended Deep Statistical Comparison

The DSC and the pDSC ranking schemes compare the performances of the optimization algorithms concerning the obtained solution values (i.e., fitness function values), while they neglect the information about where these solutions are located in the search space [27]. The information about the distribution of the obtained solutions in the search space can be used to explore the strengths and weaknesses of the compared algorithms in more detail [68].

In theory, four different scenarios (S) can be defined concerning a comparison of the obtained solutions according to their values and their distribution in the search space:

S1: There is no statistically significant difference between the compared distribution, neither with respect to the obtained solutions values nor their distribution in the search space. From here, it follows that the compared algorithms have the same exploration and exploitation capabilities.

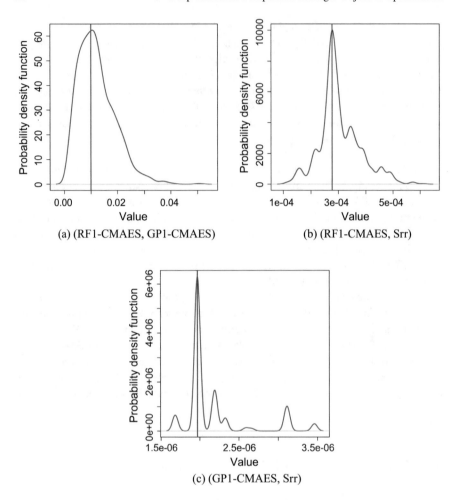

Fig. 6.9 Kernel density estimation for the obtained p-values for each pairwise comparison

S2: There is no statistical significance between the performances of the compared algorithms with regard to the distributions of the obtained solution values, but there is a statistical significance with respect to their distribution in the search space. The distribution in the search space can be sparse or clustered, which should be selected according to the user's preference. We can assume that the algorithm with the sparser distribution of obtained solutions has a better exploration capability.

S3: There is a statistical significance between the compared algorithms with respect to the distributions of the obtained solution values, but there is no statistical significance concerning their distribution in the search space. From here, it follows that the compared algorithms have the same exploration capabilities, while their exploitation capabilities differ. This result indicates that the algorithms with a lower quality of the obtained solution values should have their exploitation operators, which are included in their design, improved.

S4: There is a statistically significant difference between the algorithms' performances concerning the distributions of the obtained solution values together with their distribution in the search space. From here, it follows that the worst-performing algorithm lacks exploration capability, while its exploitation power cannot be assessed.

Since the DSC ranking scheme can test algorithms' performance concerning only the distributions of the obtained solution values, we extended it to test the algorithms' performance also considering the distributions of the obtained solutions in the search space. The extended version of the DSC ranking scheme is known as *extended Deep Statistical Comparison (eDSC)* [26].

The main difference between the DSC ranking scheme and the eDSC ranking scheme is that the DSC works with one-dimensional data (i.e., comparing the distributions of the obtained solutions values), while the eDSC looks into the distributions of the obtained solutions in the search space or handling high-dimensional data.

6.3.1 Extended Deep Statistical Comparison Ranking Scheme

Let m be the number of the compared algorithms, k the number of problem instances, n the number of runs performed by each algorithm on a single problem instance, and d the dimension of the search space (i.e., $d \geq 2$). The only difference with the DSC ranking scheme is the dimension of the search space. Each problem instance is defined as $y = f(\boldsymbol{x}) = f(x_1, x_2, \ldots, x_d)$, $f : \mathbb{R}^d \to \mathbb{R}$, where the solution value y corresponds to the point in the search space, $\boldsymbol{x} = (x_1, x_2, \ldots, x_d)$.

The eDSC ranking scheme compares the algorithms with regard to the distributions of the solutions found in the search space (i.e., $\boldsymbol{x}$). For this purpose, let us assume that $X_{i,j}$ is a $n \times d$ matrix, where $i = 1, \ldots, k$, and $j = 1, \ldots, m$. This means that for the ith problem and the jth algorithm, the rows of the matrix $X_{i,j}$ are d-dimensional vectors, which are the points in the search space that correspond to the obtained solutions values from n runs. For example, the set of matrices $\{X_{i,1}, X_{i,2}, \ldots, X_{i,m}\}$ corresponds to the data involved in the comparison performed for the ith problem. Each matrix corresponds to one algorithm that is involved in the comparison.

The idea behind the eDSC ranking scheme follows the same idea as the DSC ranking scheme, with the two key differences explained below.

First, instead of using a two-sample statistical test for comparing the distributions of the one-dimensional data, here a statistical test for comparing high-dimensional distributions should be involved. There are several options to do this available in the statistical literature. A class of consistent, asymptotic distribution free tests for the high-dimensional space is based on the nearest neighbors in the Euclidean distance metric [56, 91]. Szekely and Rizzo [100] presented a multivariate $\mathcal{E}$ test, which is universally consistent against all alternatives (not necessarily continuous) with finite second moments. The computational complexity of this test does not depend on the dimension or on the number of samples, and it is a powerful competitor with the nearest neighbor tests. The results presented in [100] suggest that the multivariate $\mathcal{E}$ test can be one of the most powerful tests available for high-dimensional data, which was also the reason to be selected and involved in the eDSC ranking scheme. So, the selected test is involved in Eq. 5.1 to make the comparison between each pair of algorithms.

All of the following steps remain the same as the DSC ranking scheme: the p-values are corrected to control the FWER, the transitivity property is checked, and then the rankings are assigned.

The second key difference with the DSC ranking scheme is based on the representative values that are selected and further involved for ranking. When the distributions of the obtained solution values by all algorithms are the same, the DSC ranking scheme ranked all the algorithms as being the same. While when the distributions are different, the ranking is made using the mean value of the multiple runs calculated for each algorithm. The algorithm that has the lowest mean value (assuming a minimization problem) is considered to be the best-performing algorithm. In the case of the eDSC ranking scheme, we cannot calculate the mean value, since high-dimensional data are involved. For this purpose, a metric that determines which search space is better must be introduced. To address this issue, the distribution type of the obtained solutions in the search space (sparse/wide or clustered/narrow) is used to find the preferred space by the user.

Let ν be a parameter that defines the desirable type of distribution for the obtained solutions in the search space. This parameter should be set in a priori, $\nu \in \{0, 1\}$, in order to define how the ranking in eDSC will be performed. When $\nu = 0$, a narrow distribution of the obtained solutions in the search space is preferred, while when $\nu = 1$ a wide distribution is preferred. To estimate this property, a measure of the multivariate spread is selected, which is the hypervolume covered by the high-dimensional distribution. This hypervolume can be calculated as a square root of the determinant of the covariance matrix [23] of $X_{i,j}$. The determinant of the covariance matrix is basically a shape factor, ranging from zero for degenerate distributions up to one when all the components are uncorrelated. It incorporates both the shape (correlation) and the size (standard deviation) information as a product of the standard deviations of the principal components [64].

Let $\Sigma_{i,j}$ be a $d \times d$ covariance matrix for the matrix $X_{i,j}$, which is a positive-definite matrix [17]. A positive-definite matrix is a symmetrical $d \times d$ matrix, S for which the scalar $z^{\mathsf{T}} S z$ is positive for every non-zero column vector $z \in \mathbb{R}^d$. Its eigenvalues are always positive. In our case, it can happen that at least one of the variables can be expressed as a linear combination of the others, so the covariance matrix is not a positive-definite matrix because it is a singular. To go beyond this, dimensionality, reduction techniques [44] should be applied to reduce the number of variables under consideration or by computing the nearest positive-definite matrix to an approximate one [57]. After performing this, we can calculate the hypervolume as

$$V_{i,j} = \sqrt{det(\Sigma_{i,j})} = \prod_{d_i=1}^{d} \sqrt{\lambda_{d_i}}, \tag{6.5}$$

where λ_{d_i} are the eigenvalues of the matrix $\Sigma_{i,j}$, obtained by using eigenvalue decomposition [45].

6.3.2 Examples

6.3.2.1 Data

To show how the eDSC ranking scheme works, the results from the Black-Box Optimization Benchmarking 2009 (BBOB 2009) competition [54] were used. BBOB 2009 is a competition that provides single-objective problems for benchmarking. The reason for selecting BBOB 2009 instead of BBOB 2015 (i.e., previously investigated for DSC and pDSC ranking schemes) is because the problem seeds that determine the location of the global optimum were the same for all the problem instances used by the algorithms involved in the competition, so a fair comparison of the distribution in the search space was possible. In later years, this was no longer the case.

Seventeen out of the 32 algorithms were selected: Cauchy-EDA, ALPS, AMAL-GAM, EDA-PSO, FULLNEWUOA, G3PCX, GA, iAMALGAM, LSfminbnd, LSstep, MCS, NELDERDOERR, POEMS, PSO, PSO_Bounds, Rosenbrock, and VNS. The selected algorithms were part of the BBOB 2009 competition. They were selected for illustrative purposes. For each algorithm, the results are available for 22 different noiseless test problems in 5 dimensionalities (2, 3, 5, 10, and 20). For each algorithm, BBOB 2009 provided data for 15 runs on each problem, which is the data we used in our experiments.

In cases when the covariance matrix is singular, the "nearPD" function from the R package "Matrix" [6] is used to compute the nearest positive-definite matrix to the covariance matrix.

Next, an example that demonstrates the benchmarking of three algorithms over the set of 22 benchmark problems will be presented where the dimension of the problems is fixed to 2, $d = 2$.

6.3.2.2 Results

Table 6.11 presents the p-values when comparing the algorithms Cauchy-EDA, MCS, and iAMALGAM using the set of 22 problems. The p_{value_Y} corresponds to the p-value when comparing the obtained solution values using the Friedman test (analyzed with the DSC ranking scheme), while the p_{value_X} corresponds to the p-value when comparing the distributions of the obtained solutions in the search space using the Friedman test (analyzed with the eDSC ranking scheme).

Table 6.11 Statistical comparisons of Cauchy-EDA, MCS, and iAMALGAM

Algorithms	p_{valueY}	p_{valueX}
Cauchy-EDA, MCS, iAMALGAM	(0.44)	(0.95)

p_{valueY} corresponds to the p-value when comparing the obtained solutions values using the Friedman test

p_{valueX} corresponds to the p-value when comparing the distributions of the obtained solutions in search space using the Friedman test

Using the p-values presented, it follows that there is no statistical significance between the performance of the algorithms over the set of 22 problems considering the obtained solution values. The same results are also valid when the algorithms are compared considering the distributions of the obtained solutions in the search space. However, to explore this example in more detail, we presented the rankings obtained by the DSC ranking scheme (using the solution values) and the eDSC ranking scheme (using the distribution of the solutions in the search space) for each benchmark problem, separately. The DSC and eDSC rankings are presented in Table 6.12.

To see what happens at the single problem level, let us focus on the 5th, 7th, and 20th problems (Table 6.12). The DSC rankings obtained on the f_5 problem are 2.00, 2.00, and 2.00. Using the DSC ranking scheme, the algorithms are compared according to the obtained solution values. All the algorithms have a zero fitness value in all 15 runs. From here, it follows that they have the same distribution of the obtained solution values (see also Fig. 6.10). This is also checked with the two-sample AD test, which provides a p-value of 1.00 for all the involved pairwise comparisons. After correcting the p-values with the Bonferroni correction, the transitivity of the p-value matrix is satisfied, so all the algorithms belong to one set and they receive the same ranking.

Next, for the same problem, f_5, the obtained eDSC rankings are 3.00, 1.00, and 2.00. In Fig. 6.11, for each of the three algorithms, the contour plots of the probability density functions of the obtained two-dimensional solutions are presented. Using the contour plots for the three algorithms, it is difficult to compare them. This happens because they all have their own color scale and different values on the x_1 and x_2 axes. However, this is for a better visualization, since if the same color scale was used (i.e., the one used in Fig. 6.11b)), then both the contour plots presented in Fig. 6.11a, c would be blue. With regard to the axis values, if we selected the same values on the x_1 and x_2 axes (the ones used in Fig. 6.11a), the contour plots presented in Fig. 6.11b, c would be too narrow and the shape of the distribution would not be visible.

From Fig. 6.11, it is obvious that the shapes of the distributions and their spreads in the search space are different. This indicates that the distributions of the solutions in the search space differ. Furthermore, this was checked with the multivariate $\mathcal{E}$ test. The obtained p-values for all the pairwise comparisons are 0.01 (Cauchy-EDA, MCS), 0.26 (Cauchy-EDA, iAMALGAM), and 0.22 (MCS, iAMALGAM), which are further corrected using the Bonferroni correction. The transitivity of the p-value

Table 6.12 Rankings for the algorithms A_1 = Cauchy-EDA, A_2 = MCS, and A_3 = iAMALGAM

F	DSC			eDSC		
	A_1	A_2	A_3	A_1	A_2	A_3
f_1	2.50	1.00	2.50	2.00	2.00	2.00
f_2	1.50	3.00	1.50	2.00	2.00	2.00
f_3	2.50	1.00	2.50	2.00	2.00	2.00
f_4	3.00	1.50	1.50	2.00	2.00	2.00
f_5	**2.00**	**2.00**	**2.00**	**3.00**	**1.00**	**2.00**
f_6	2.00	2.00	2.00	2.00	2.00	2.00
f_7	**2.00**	**2.00**	**2.00**	**2.00**	**2.00**	**2.00**
f_8	2.50	1.00	2.50	2.00	2.00	2.00
f_9	2.50	1.00	2.50	2.00	2.00	2.00
f_{10}	1.50	3.00	1.50	2.00	2.00	2.00
f_{11}	1.50	3.00	1.50	2.00	2.00	2.00
f_{12}	2.00	2.00	2.00	2.00	2.00	2.00
f_{13}	1.50	3.00	1.50	2.00	2.00	2.00
f_{14}	1.50	3.00	1.50	2.00	2.00	2.00
f_{15}	2.00	2.00	2.00	2.00	2.00	2.00
f_{16}	1.50	3.00	1.50	2.00	2.00	2.00
f_{17}	2.50	1.00	2.50	2.00	2.00	2.00
f_{18}	2.50	1.00	2.50	2.00	2.00	2.00
f_{19}	3.00	1.50	1.50	2.00	2.00	2.00
f_{20}	**3.00**	**1.00**	**2.00**	**2.00**	**2.00**	**2.00**
f_{21}	2.50	1.00	2.50	2.00	2.00	2.00
f_{22}	3.00	1.00	2.00	2.00	2.00	2.00

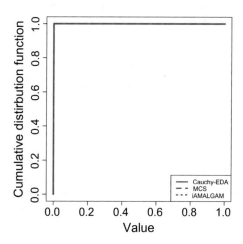

Fig. 6.10 Empirical cumulative distribution functions for f_5

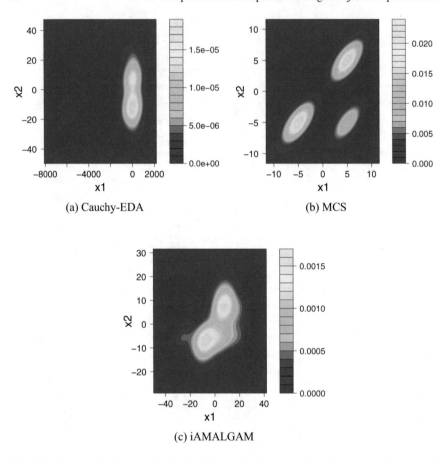

Fig. 6.11 Contour plots for empirical probability density functions of the obtained two-dimensional solutions for each algorithm on f_5

matrix is not satisfied, so all the algorithms should receive different rankings. For this purpose, the hypervolume covered by the distribution needs to be calculated as a measure of the multivariate spread. To find it, the covariance matrix must be calculated from the matrix where the points of the obtained solutions in the search space are stored. This should be done separately for each algorithm, $X_{5,j}$, $j = 1, \ldots, 3$. Furthermore, applying eigenvalue decomposition to the covariance matrix results in its eigenvalues being needed to calculate the hypervolume covered by each distribution. The hypervolumes are $20, 775.698$ (Cauchy-EDA), 19.166 (MCS), and 84.063 (iAMALGAM). The eDSC ranking scheme has one additional parameter, ν, which should be set by the user, depending on whether we are interested in a narrow or a wide distribution. Let us assume that $\nu = 0$, so we are interested in a narrow search space, which means that a smaller hypervolume value should be ranked as better. Using this information, the obtained eDSC rankings are 3.00, 1.00 and 2.00. If we

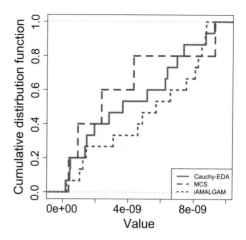

Fig. 6.12 Empirical cumulative distribution functions for f_7

are interested in a wide distribution, $\nu = 1$, the eDSC ranking will only be reordered where the higher value is the better one (i.e., 1.00, 3.00, and 2.00).

Using the information obtained by the DSC and eDSC ranking schemes on f_5, we can conclude that the three compared algorithms are not statistically significant according to the comparison made using the obtained solution values, but there is a statistical significance between them according to the comparison made with respect to the distribution of the solutions in the search space. This result indicates that even if the algorithms have different exploration capabilities, they are all able to find good solutions (i.e., to have the same exploitation capabilities).

Looking at the 7th problem, f_7, the DSC rankings are 2.00, 2.00, and 2.00. Figure 6.12 presents the cumulative distribution functions for the solution values. To check the statistical significance between the distributions, the two-sample AD test was used. The obtained p-values are 0.39 (Cauchy-EDA, MCS), 0.64 (Cauchy-EDA, iAMALGAM), and 0.05 (MCS, iAMALGAM). Furthermore, these p-values are corrected using the Bonferroni correction, the transitivity of the p-values matrix is satisfied, and all the algorithms belong to one set, {Cauchy-EDA, MCS, iAMAL-GAM}, so they should achieve the same ranking.

With regard to the comparison made that considers the distribution of the solutions in the search space on f_7, the eDSC rankings are 2.00, 2.00, and 2.00. Figure 6.13 presents the contour plots for the probability density functions of the two-dimensional solutions for each algorithm for f_7.

From the figure, it is clear that there is no statistical significance between the algorithms considering the distributions of the solutions in the search space. This is also checked by the multivariate $\mathcal{E}$ test and the obtained p-values are 0.98 (Cauchy-EDA, MCS), 0.99 (Cauchy-EDA, iAMALGAM), and 0.98 (MCS, iAMALGAM), which are further corrected using the Bonferroni correction. The transitivity of the p-value matrix is satisfied, all the algorithms belong to the same set, so they achieve the same ranking.

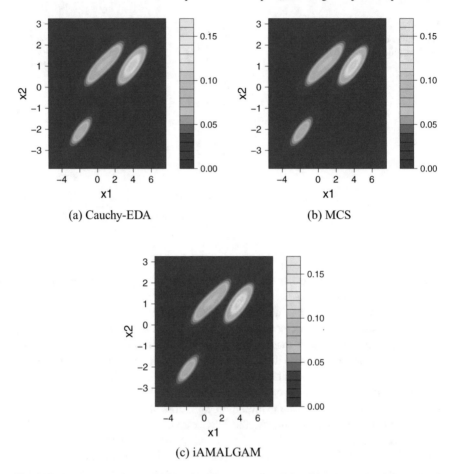

Fig. 6.13 Contour plots for empirical probability density functions of the obtained two-dimensional solutions for each algorithm on f_7

Using the results for the DSC and the eDSC ranking schemes on f_7, it follows that all the algorithms have the same exploration and exploitation capabilities.

Next, for the 20th problem, the DSC rankings are 3.00, 1.00, and 2.00. Figure 6.14 presents the cumulative distribution functions for the obtained solution values for f_{20}. There is a statistical significance between the cumulative distribution functions of Cauchy-EDA, MCS, and iAMALGAM. This is also proved using the two-sample AD test that provides the p-values: 0.00 (Cauchy-EDA, MCS), 0.00 (Cauchy-EDA, iAMALGAM), and 0.00 (MCS, iAMALGAM).

The eDSC rankings for f_{20} are 2.00, 2.00, and 2.00. The contour plots of the probability density functions of the obtained two-dimensional data for each algorithm show no statistical significance between the distributions of the solutions in the search

Fig. 6.14 Empirical
cumulative distribution
functions for f_{20}

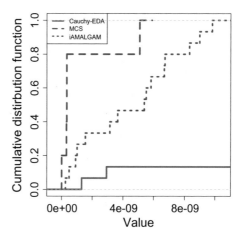

space (see Fig. 6.15). The multivariate $\mathcal{E}$ test also provides the same result. The
obtained p-values for the pairwise comparisons are 0.88 (Cauchy-EDA, MCS), 0.88
(Cauchy-EDA, iAMALGAM), and 0.88 (MCS, iAMALGAM).

When the comparison is made on the f_{20} problem, there is a statistical significance
between the distributions for the obtained solutions values, but there is no statistical
significance between the algorithms according to the distributions of the obtained
solutions in the search space. This result indicates that all three algorithms are able
to find the same region with good solutions (i.e., the same exploration capabilities),
but one of them is able to find statistically better solutions than the other two from
the same region (i.e., different exploitation capabilities). From here, it follows that
the MSC algorithm has a better exploitation power than the other two algorithms.

6.4   Summary

In this chapter:

- We presented examples of how the Deep Statistical Comparison approach is used
  for testing whether there is a statistical significance between the performances of
  single-objective optimization algorithms.
- We introduced a practical Deep Statistical Comparison approach and we presented
  examples to point out the difference that exists between statistical and practical significance in benchmarking studies. This analysis can help us to detect whether the
  statistical significance that can be found between the performance of the compared
  algorithms is also relevant for real-world applications.

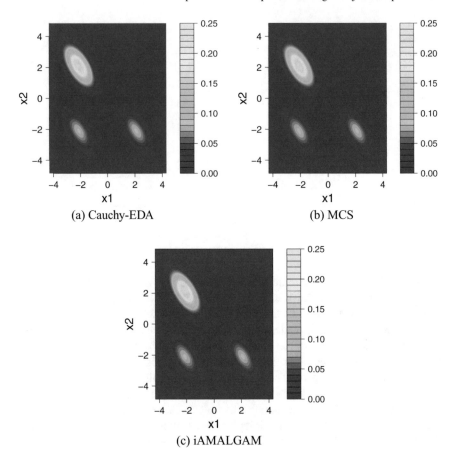

Fig. 6.15 Contour plots for empirical probability density functions of the obtained two-dimensional solutions for each algorithm on f_{20}

- We introduced an extended Deep Statistical Comparison approach that ranks the algorithms based on the distribution of the solutions in the search space. It uses high-dimensional experimental data, which are the location of the obtained solutions in the search space. This analysis allows us to explore the strengths and weaknesses of the compared algorithms, providing more details about their exploration and exploitation capabilities.
- All the examples presented in this chapter can be repeated and reproduced using the code presented in Chap. 8.

Chapter 7
Deep Statistical Comparison in Multi-Objective Optimization

7.1 Single-Quality-Indicator Analysis

As already mentioned in Sect. 2.2.2, the performance assessment of a multi-objective optimization algorithm can be done using quality indicators, which are mathematical formulas that map the high-dimensional approximation set into a scalar number. By using them, the high-dimensional data that is the result of the experiments is transformed into one-dimensional data. Performing this transformation allows us to utilize the Deep Statistical Comparison ranking scheme [36]. The main difference with the DSC ranking scheme used in the single-objective optimization is the type of data involved in the comparison. In single-objective optimization, we are comparing the distributions of the obtained solution values. In multi-objective optimization, first, a quality indicator should be selected by the user to transform the approximation sets from the multiple independent runs of each algorithm on each problem into scalar numbers. Further, the quality-indicator data is involved in the comparison.

We would like to point here that the DSC ranking scheme assumes lower values to be better (i.e., typically interested in a minimization problem). However, depending on the definition of the quality indicator, it can happen that greater values are better. In such cases, before using the DSC ranking scheme, the quality-indicator data should be first transformed by multiplying it with -1.

7.1.1 Examples

7.1.1.1 Data

The data involved in the examples is publicly available and can be accessed in [102]. The obtained results from six multi-objective optimization algorithms are available for 16 multi-objective optimization test problems. Three out of six algorithms

© The Author(s), under exclusive license to Springer Nature Switzerland AG 2022
T. Eftimov and P. Korošec, *Deep Statistical Comparison for Meta-heuristic Stochastic Optimization Algorithms*, Natural Computing Series,
https://doi.org/10.1007/978-3-030-96917-2_7

(NSGA-II, SPEA2, and IBEA) are genetic algorithms, while the remaining three are their differential evolution-based variants $DEMO^{NS-II}$, $DEMO^{SP2}$, and $DEMO^{IB}$. The sixteen test problems include seven DLTZ [19] and nine WFG problems [61]. The results from running the algorithms are available when the test problems are investigated for a different number of objectives (2, 3, and 4). All test problems assume the minimization of all objectives. The tested hyper-parameters used for the algorithms are available in [102]. Each algorithm was run 30 times on each optimization problem. Before calculating the quality indicators, each approximated *Pareto* front was normalized.

To compare the algorithms, four quality indicators are selected: hypervolume, epsilon indicator, r_2 indicator, and generational distance. For the hypervolume calculation, the reference point $(1,\ldots,1)$ was used, while for the other quality indicators, the reference set consisted of non-dominated solutions acquired from all runs of each algorithm on a given problem.

The DSC ranking scheme is used with the two-sample AD test. The benefits of using it are presented in Sect. 6.1.2.2. The significance level for it is set to 0.05. Bonferroni correction is used to correct the p-values used in the DSC ranking scheme.

7.1.1.2   Results

Here, an example of multiple comparisons among algorithms is presented. The comparison is done between three algorithms ($DEMO^{SP2}$, $DEMO^{NS-II}$, and NSGA-II). For each of the sixteen problems, the algorithms are compared using the DSC ranking scheme. Further, the DSC rankings are compared with the rankings obtained when the fractional ranking scheme (i.e., the ranking scheme used by the Friedman test) is applied using the mean of the quality-indicator data of an algorithm over 30 runs on a given problem as a representative value for this algorithm on that problem. The fractional ranking scheme is most commonly used when the data is compared in a paired multiple-problem scenario.

Table 7.1 presents the DSC rankings obtained for the three algorithms for each problem separately, while Table 7.2 presents the rankings obtained by the fractional ranking scheme.

Focusing on the comparison done with regard to the hypervolume, the rankings obtained with the DSC ranking scheme and the fractional ranking scheme differ for four problems: DLTZ3, DTLZ5, WFG8, and WFG9. The rankings with regard to the r_2 indicator differ in seven problems: DTLZ4, DTLZ5, WFG2, WFG3, WFG4, WFG5, and WFG9. The rankings according to the epsilon indicator differ in 8 problems: DTLZ3, DTLZ5, WFG2, WFG3, WFG6, WFG7, WFG8, and WFG9, while the rankings concerning the generational distance indicator differ in four problems: DTLZ3, WFG2, WFG3, and WFG6. Similar to the single-objective optimization (see Sect. 6.1.2.2), this happens because the difference between the calculated mean values can be in some small ϵ-neighborhood, so irrespective of that, different rankings will be assigned by the fractional ranking scheme. However, when the DSC ranking scheme is used, if the quality-indicator distributions of the algorithms are the same,

Table 7.1 DSC rankings for each quality indicator of the algorithms, A_1=DEMO^{SP2}, A_2= DEMO^{NS-II}, and A_3=NSGA-II

F	Hypervolume			r_2			Epsilon			Generational distance		
	A_1	A_2	A_3	A_1	A_2	A_3	A_1	A_2	A_3	A_1	A_2	A_3
DTLZ1	2.00	1.00	3.00	1.00	2.00	3.00	1.00	2.00	3.00	1.00	2.00	3.00
DTLZ2	2.00	1.00	3.00	3.00	1.00	2.00	2.00	1.00	3.00	2.00	1.00	3.00
DTLZ3	1.50	1.50	3.00	2.00	1.00	3.00	1.50	1.50	3.00	1.50	1.50	3.00
DTLZ4	1.00	2.00	3.00	1.00	2.50	2.50	1.00	2.00	3.00	1.00	2.00	3.00
DTLZ5	2.50	2.50	1.00	1.50	1.50	3.00	2.00	2.00	2.00	1.00	3.00	2.00
DTLZ6	2.00	1.00	3.00	2.00	1.00	3.00	2.00	1.00	3.00	1.00	2.00	3.00
DTLZ7	2.00	1.00	3.00	2.00	1.00	3.00	2.00	1.00	3.00	2.00	1.00	3.00
WFG1	1.00	2.00	3.00	1.00	2.00	3.00	1.00	2.00	3.00	1.00	3.00	2.00
WFG2	1.00	2.00	3.00	1.00	2.50	2.50	1.00	2.50	2.50	1.50	3.00	1.50
WFG3	1.00	3.00	2.00	1.00	2.50	2.50	1.00	2.50	2.50	1.00	2.50	2.50
WFG4	1.00	2.00	3.00	2.50	1.00	2.50	2.00	1.00	3.00	3.00	2.00	1.00
WFG5	3.00	2.00	1.00	3.00	1.50	1.50	1.00	3.00	2.00	3.00	2.00	1.00
WFG6	1.00	2.00	3.00	2.00	1.00	3.00	1.00	2.50	2.50	3.00	1.50	1.50
WFG7	1.00	2.00	3.00	2.00	1.00	3.00	1.00	2.50	2.50	3.00	2.00	1.00
WFG8	1.00	2.50	2.50	1.00	2.00	3.00	1.00	2.50	2.50	1.00	3.00	2.00
WFG9	1.00	2.50	2.50	1.50	1.50	3.00	**1.00**	**2.50**	**2.50**	3.00	2.00	1.00

Table 7.2 Fractional rankings for each quality indicator of the algorithms, A_1=DEMO^{SP2}, A_2= DEMO^{NS-II}, and A_3=NSGA-II, obtained on averages of each quality indicator over 30 runs

F	Hypervolume			r_2			Epsilon			Generational distance		
	A_1	A_2	A_3	A_1	A_2	A_3	A_1	A_2	A_3	A_1	A_2	A_3
DTLZ1	2.00	1.00	3.00	1.00	2.00	3.00	1.00	2.00	3.00	1.00	2.00	3.00
DTLZ2	2.00	1.00	3.00	3.00	1.00	2.00	2.00	1.00	3.00	2.00	1.00	3.00
DTLZ3	1.00	2.00	3.00	2.00	1.00	3.00	1.00	2.00	3.00	1.00	2.00	3.00
DTLZ4	1.00	2.00	3.00	1.00	2.00	3.00	1.00	2.00	3.00	1.00	2.00	3.00
DTLZ5	3.00	2.00	1.00	2.00	1.00	3.00	1.00	2.00	3.00	1.00	3.00	2.00
DTLZ6	2.00	1.00	3.00	2.00	1.00	3.00	2.00	1.00	3.00	1.00	2.00	3.00
DTLZ7	2.00	1.00	3.00	2.00	1.00	3.00	2.00	1.00	3.00	2.00	1.00	3.00
WFG1	1.00	2.00	3.00	1.00	2.00	3.00	1.00	2.00	3.00	1.00	3.00	2.00
WFG2	1.00	2.00	3.00	1.00	2.00	3.00	1.00	2.00	3.00	1.00	3.00	2.00
WFG3	1.00	3.00	2.00	1.00	3.00	2.00	1.00	2.00	3.00	1.00	3.00	2.00
WFG4	1.00	2.00	3.00	2.00	1.00	3.00	2.00	1.00	3.00	3.00	2.00	1.00
WFG5	3.00	2.00	1.00	3.00	1.00	2.00	1.00	3.00	2.00	3.00	2.00	1.00
WFG6	1.00	2.00	3.00	2.00	1.00	3.00	1.00	2.00	3.00	3.00	2.00	1.00
WFG7	1.00	2.00	3.00	2.00	1.00	3.00	1.00	2.00	3.00	3.00	2.00	1.00
WFG8	1.00	3.00	2.00	1.00	2.00	3.00	1.00	3.00	2.00	1.00	3.00	2.00
WFG9	1.00	3.00	2.00	1.00	2.00	3.00	**1.00**	**2.00**	**3.00**	3.00	2.00	1.00

they obtain the same ranking. So, the rankings are obtained according to the whole distribution and not relying only on one statistic, which is the mean in our example.

To clarify the differences that exist between the DSC rankings and the rankings obtained by the fractional ranking scheme, Fig. 7.1 presents the cumulative distribution functions (i.e., the step functions) and the mean values (i.e., the horizontal lines) for the epsilon indicator calculated for the algorithms on WFG9.

Using the figure, it seems that there is no difference between the cumulative distribution functions of the epsilon indicator for the algorithms DEMO^{NS-II} and NSGA-II. However, it seems that both distributions differ from the cumulative distribution of the DEMO^{SP2}. This is further checked with the two-sample AD test and the steps used in the DSC ranking scheme. The p-values obtained for each pairwise comparison are 0.00 (DEMO^{SP2}, DEMO^{NS-II}), 0.00 (DEMO^{SP2}, NSGA-II), and 0.61 (DEMO^{NS-II}, NSGA-II). Further, by correcting them using the Bonferroni correction, it follows that the transitivity of the matrix used in the DSC ranking scheme is satisfied. Then, the set of all algorithms is split into two disjoint sets {DEMO^{SP2}} and {DEMO^{NS-II}, NSGA-II} and the algorithms are ranked 1.00, 2.50, and 2.50, respectively. By using the fractional ranking scheme, it is obvious that the averages of the epsilon indicator

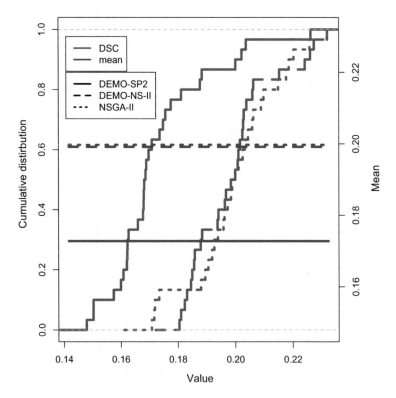

Fig. 7.1 Empirical cumulative distribution functions (step functions) and the mean values (horizontal lines) of the epsilon quality indicator for WFG9 problem

for DEMO$^{\mathrm{NS-II}}$ and NSGA-II are in some small ϵ-neighborhood, so the algorithms are ranked 1.00, 2.00, and 3.00. This further can affect the test statistic used by the selected omnibus statistical test, so the statistical outcome will be questionable.

Next, using the Friedman test as an appropriate omnibus statistical test, we can find the statistical results for the scenario where all problems are involved. If we focus on the hypervolume and the epsilon indicators in combination with both ranking schemes, the result from the Friedman test are p-values around 0.00. It follows that there is a significant difference between the performance of the compared algorithms over the multiple problems that are used in the comparison, no matter which ranking scheme is used. In this example, in the case of multiple-problem analysis, there is no difference between the result obtained by both ranking schemes, however, there are differences on a single-problem level. In general, the differences that exist on a single-problem level can influence the result for multiple-problem analysis.

In cases when we are interested in multiple comparisons with a control algorithm, the statistical scenario has been already explained in Sect. 6.1.2.3. The main difference is that instead of using the obtained solution values from the single-objective optimization, the quality-indicator data will be involved in the comparison.

7.2 Ensemble of Quality Indicator Analysis

Even though there are some insights into algorithms' performance by performing performance assessment using quality indicators, it is well known that the selection of a quality indicator(s) has an impact on the benchmarking conclusions [69]. Different quality indicators provide different explanations and results. To address this challenge and to make some more general conclusions, we proposed three different ensemble heuristics for making a statistical comparison of multi-objective optimization algorithms. Each ensemble heuristic uses a set of quality indicators specified by the user and further utilizes the DSC ranking scheme. The three proposed heuristics are an *average ensemble*, a *hierarchical majority vote ensemble*, and a *data-driven ensemble*.

Let $A = \{A_1, A_2, \ldots, A_m\}$ be the set of algorithms that are compared using the set of quality indicators $Q = \{q_1, q_2, \ldots, q_n\}$, where m is the number of algorithms and n the number of quality indicators involved in the comparison. Let us define an $m \times n$ matrix (see Table 7.3) that contains the DSC rankings obtained for the algorithms for each quality indicator separately.

The column vector $\boldsymbol{q}_j = \left[q_j(A_1), q_j(A_2), \ldots, q_j(A_m)\right]^{\mathrm{T}}$, where $j = 1, \ldots, n$, corresponds to the DSC rankings obtained when comparing the algorithms using the data obtained for the jth quality indicator.

The DSC ranking scheme ranks the best algorithm as 1.00, the second best as 2.00, and so on. Lower DSC ranking values are preferable. For example, let us assume that three algorithms are compared on a given multi-objective optimization problem using a single-quality indicator. The obtained DSC rankings of such a comparison are 1.50,

Table 7.3 Matrix with DSC rankings

A	q_1	q_2	$\ldots$	q_n
A_1	$q_1(A_1)$	$q_2(A_1)$	$\ldots$	$q_n(A_1)$
A_2	$q_1(A_2)$	$q_2(A_2)$	$\ldots$	$q_n(A_2)$
$\vdots$	$\vdots$	$\vdots$	$\vdots$	
A_m	$q_1(A_m)$	$q_2(A_m)$	$\ldots$	$q_n(A_m)$

3.00, and 1.50. This result indicates that the first and the third algorithm perform equally and are better than the second algorithm. The idea behind all three ensemble heuristics is to combine the DSC rankings obtained by each quality indicator for each algorithm for a given problem. By doing this, it does not matter if the DSC rankings are 1.50, 3.00, and 1.50 or 1.00, 3.00, and 1.00 because having 1.00 or 1.50 means that the algorithm is the best regarding some quality indicator. The DSC rankings scheme cannot provide 1.00, 3.00, and 1.00, since it follows the idea of a fractional ranking scheme. In order to transform them, the DSC rankings obtained for each quality indicator must be transformed using a *standard competition ranking scheme*. The standard competition ranking scheme is the adopted ranking scheme from the literature used for competitions. Using it, items that compare equally receive the same ranking with a gap left in the rankings. The number of rankings that are left out in this gap is one less than the number of items that are compared equally. Each item ranking is one plus the number of items ranked above it. Using the standard competition ranking, it means that when two (or more) competitors tie for a position in the ranking, the position of all those ranked below them is unaffected. Before the ensemble heuristics are defined, each column vector q_j should be transformed using the standard competition ranking scheme.

7.2.1 Average Ensemble

The average ensemble heuristic is defined as an average of the transformed DSC rankings [35]. The ranking for each algorithm on each problem is the average of its transformed DSC rankings obtained for all quality indicators for that problem. The algorithm with the lowest ranking is the best one. One weakness of this heuristic is its sensitivity when outliers are presented. To avoid this, instead of average, the median of the DSC rankings can also be used. This ensemble can be used to provide a more general conclusion in a benchmarking study.

7.2.2 Hierarchical Majority Vote Ensemble

The hierarchical majority vote ensemble heuristic counts which algorithm is ranked with the best transformed DSC ranking in the most number of quality indicators [35]. If the winner is only one algorithm, it will be ranked as 1. Further, the remaining algorithms are again involved in the comparison, comparing them again starting from the best transformed DSC ranking. If there exists more than one algorithm with the same number of wins, then these algorithms are compared concerning the next ranking, or which of them is more times ranked with the next transformed DSC ranking. This is recursively repeated until all of the algorithms obtain their rankings. This ensemble is recommended when dynamic multi-objective optimization is performed, where the performance can be treated as counting wins and losses.

7.2.3 Data-Driven Ensemble

The average and the hierarchical majority vote ensembles treat equally all quality indicators that are specified by the user. However, the statistical results of such comparisons can be affected by the user preference in the selection of the quality indicators. If we have more correlated quality indicators in favor of some algorithm, then this indirectly associates that the algorithm will be superior to the other algorithms involved in the comparison. To go beyond this, a data-driven ensemble heuristic is proposed, which combines the DSC rankings of the quality indicators using the preference of each quality indicator estimated by its entropy [38].

This ensemble utilizes the PROMETHEE II method [12]. PROMETHEE methods are used in decision-making to solve a decision problem in which a set of alternatives are evaluated according to a set of criteria that are often conflicting. The idea behind this ensemble heuristic is based on the information that is presented in all pairwise comparisons between all algorithms for each quality indicator. The differences between the DSC rankings for each pair of algorithms with regard to a given quality indicator are taken into consideration. Before the ranking is performed, a preference function should be defined for each quality indicator. It is defined as the degree of preference of algorithm A_1 over algorithm A_2:

$$P_j(A_1, A_2) = \begin{cases} p_j(d_j(A_1, A_2)), & \text{if maximizing the quality indicator} \\ p_j(-d_j(A_1, A_2)), & \text{if minimizing the quality indicator} \end{cases}, \quad (7.1)$$

where $d_j(A_1, A_2) = q_j(A_1) - q_j(A_2)$ is the difference between the DSC rankings of the algorithms for the quality indicator, q_j, and $p_j(\cdot)$ is a generalized preference function assigned to the quality indicator. There exist six types of generalized preference functions [12]. In our case, the usual preference function is selected and used for each quality indicator. The usual preference function is presented as follows:

$$p(x) = \begin{cases} 0, & x \le 0 \\ 1, & x > 0 \end{cases}. \tag{7.2}$$

Next, the average preference index and outranking flows should be calculated. The average preference index for each pair of algorithms gives information about the global comparison between them using all quality indicators. The average preference index can be calculated as

$$\pi(A_1, A_2) = \frac{1}{n} \sum_{j=1}^{n} w_j P_j(A_1, A_2), \tag{7.3}$$

where w_j represents the relative significance (i.e., weight) of the jth quality indicator. The higher weight means higher importance in the ranking process. The average preference index is not a symmetric function, $\pi(A_1, A_2) \ne \pi(A_2, A_1)$.

The PROMETHEE II method is highly dependent on the selection of the weights. To calculate the weights, we used the Shannon entropy weight method (explained in Sect. 7.2.3.1).

The outranking flow involves calculating the positive preference flow, the negative preference flow, and the net flow. The ranking is further performed using the net flow calculated for each algorithm. The net flow is the difference between the positive preference flow, $\phi(A_i^+)$, and the negative preference flow of the algorithm, $\phi(A_i^-)$. The positive preference flow gives information about how a given algorithm is globally better than the other algorithms, while the negative preference flow gives information about how a given algorithm is outranked by all the other algorithms. The positive and the negative preference flows are defined as

$$\phi(A_i^+) = \frac{1}{(n-1)} \sum_{x \in A} \pi(A_i, x),$$

$$\phi(A_i^-) = \frac{1}{(n-1)} \sum_{x \in A} \pi(x, A_i). \tag{7.4}$$

The net flow of an algorithm is defined as

$$\phi(A_i) = \phi(A_i^+) - \phi(A_i^-). \tag{7.5}$$

The PROMETHEE II method ranks the algorithms by ordering them according to decreasing values of net flows.

7.2.3.1 The Shannon Entropy Weighted Method

The Shannon entropy weighted method [11] can be used to calculate the weights of the quality indicators involved in the study. For this purpose, the matrix presented in

Table 7.3 needs to be normalized. Because the smaller value is preferred, the matrix is normalized using the following equation:

$$q_j(A_i)' = \frac{\max_i(q_j(A_i)) - q_j(A_i)}{\max_i(q_j(A_i)) - \min_i(q_j(A_i))}, \tag{7.6}$$

where $q_j(A_i)'$ is the normalized value for $q_j(A_i)$.

The entropy for each quality indicator is defined as

$$e_j = K \sum_{i=1}^{m} W\left(\frac{q_j(A_i)'}{D_j}\right), \tag{7.7}$$

where D_j is the sum of the jth quality indicator in all algorithms, $D_j = \sum_{i=1}^{m} q_j(A_i)'$, K is the normalized coefficient, $K = \frac{1}{(e^{0.5}-1)m}$, and W is a function defined as $W(x) = xe^{(1-x)} + (1-x)e^x - 1$.

The weight of each quality indicator used in Eq. 7.3 is calculated using the following equation:

$$w_j = \frac{\frac{1}{(n-E)}(1 - e_j)}{\sum_{j=1}^{n}\left[\frac{1}{(n-E)}(1 - e_j)\right]}, \tag{7.8}$$

where E is the sum of entropies, $E = \sum_{j=1}^{n} e_j$.

7.2.4 Examples

To show how the ensemble heuristics work, we consider the same example presented in Sect. 7.1.1.2, where the comparison between DEMO^{SP2}, DEMO^{NS-II}, and NSGA-II is presented. The set of quality indicators that are utilized by the ensembles are hypervolume, r_2, epsilon, and generational distance marked as q_1, q_2, q_3, and q_4, respectively. The DSC rankings for each algorithm on each problem are presented in Table 7.1 for each quality indicator separately. For the purpose of the ensembles, these rankings are further transformed with the standard competition ranking scheme (see Table 7.4).

Table 7.5 presents the rankings that are obtained for each algorithm on each problem using the three ensemble heuristics. From it, we can see that the rankings obtained using data-driven ensemble with DSC differ from the rankings obtained using the average ensemble with DSC or the hierarchical majority vote with DSC only in two bolded problems: DTLZ5 and WFG7.

To understand how the ranking is performed using the different ensemble heuristics, let us focus on explaining them on a single-problem level. We are not going into more details about the average heuristic, since using it the average value from

Table 7.4 Transformed DSC rankings for each quality indicator of the algorithms, A_1=DEMO^{SP2}, A_2=DEMO^{NS-II}, and A_3=NSGA-II

F	Hypervolume			r_2			Epsilon			Generational distance		
	A_1	A_2	A_3	A_1	A_2	A_3	A_1	A_2	A_3	A_1	A_2	A_3
DTLZ1	2.00	1.00	3.00	1.00	2.00	3.00	1.00	2.00	3.00	1.00	2.00	3.00
DTLZ2	2.00	1.00	3.00	3.00	1.00	2.00	2.00	1.00	3.00	2.00	1.00	3.00
DTLZ3	1.00	1.00	3.00	2.00	1.00	3.00	1.00	1.00	3.00	1.00	1.00	3.00
DTLZ4	1.00	2.00	3.00	1.00	2.00	2.00	1.00	2.00	3.00	1.00	2.00	3.00
DTLZ5	2.00	2.00	1.00	1.00	1.00	3.00	1.00	1.00	1.00	1.00	3.00	2.00
DTLZ6	2.00	1.00	3.00	2.00	1.00	3.00	2.00	1.00	3.00	1.00	2.00	3.00
DTLZ7	2.00	1.00	3.00	2.00	1.00	3.00	2.00	1.00	3.00	2.00	1.00	3.00
WFG1	1.00	2.00	3.00	1.00	2.00	3.00	1.00	2.00	3.00	1.00	3.00	2.00
WFG2	1.00	2.00	3.00	1.00	2.00	2.00	1.00	2.00	2.00	1.00	3.00	1.00
WFG3	1.00	3.00	2.00	1.00	2.00	2.00	1.00	2.00	2.00	1.00	2.00	2.00
WFG4	1.00	2.00	3.00	2.00	1.00	2.00	2.00	1.00	3.00	3.00	2.00	1.00
WFG5	3.00	2.00	1.00	3.00	1.00	1.00	1.00	3.00	2.00	3.00	2.00	1.00
WFG6	1.00	2.00	3.00	2.00	1.00	3.00	1.00	2.00	2.00	3.00	1.00	1.00
WFG7	1.00	2.00	3.00	2.00	1.00	3.00	1.00	2.00	2.00	3.00	2.00	1.00
WFG8	1.00	2.00	2.00	1.00	2.00	3.00	1.00	2.00	2.00	1.00	3.00	2.00
WFG9	1.00	2.00	2.00	1.00	1.00	3.00	1.00	2.00	2.00	3.00	2.00	1.00

Table 7.5 Ensemble combiner for the algorithms: A_1=DEMO^{SP2}, A_2=DEMO^{NS-II}, and A_3= NSGA-II

F	Average			Hierarchical			Data-driven		
	A_1	A_2	A_3	A_1	A_2	A_3	A_1	A_2	A_3
DTLZ1	1.00	2.00	3.00	1.00	2.00	3.00	1.00	2.00	3.00
DTLZ2	2.00	1.00	3.00	2.00	1.00	3.00	2.00	1.00	3.00
DTLZ3	2.00	1.00	3.00	2.00	1.00	3.00	2.00	1.00	3.00
DTLZ4	1.00	2.00	3.00	1.00	2.00	3.00	1.00	2.00	3.00
DTLZ5	**1.00**	**2.50**	**2.50**	**1.00**	**2.50**	**2.50**	**2.00**	**3.00**	**1.00**
DTLZ6	2.00	1.00	3.00	2.00	1.00	3.00	2.00	1.00	3.00
DTLZ7	2.00	1.00	3.00	2.00	1.00	3.00	2.00	1.00	3.00
WFG1	1.00	2.00	3.00	1.00	2.00	3.00	1.00	2.00	3.00
WFG2	1.00	3.00	2.00	1.00	3.00	2.00	1.00	3.00	2.00
WFG3	1.00	3.00	2.00	1.00	3.00	2.00	1.00	3.00	2.00
WFG4	2.00	1.00	3.00	2.00	1.00	3.00	2.00	1.00	3.00
WFG5	3.00	2.00	1.00	3.00	2.00	1.00	3.00	2.00	1.00
WFG6	2.00	1.00	3.00	2.00	1.00	3.00	1.00	2.00	3.00
WFG7	**1.50**	**1.50**	**3.00**	**1.00**	**2.00**	**3.00**	**1.00**	**2.00**	**3.00**
WFG8	1.00	2.50	2.50	1.00	2.50	2.50	1.00	2.50	2.50
WFG9	1.00	2.00	3.00	1.00	2.00	3.00	1.00	2.00	3.00

Table 7.6 Hierarchical majority vote for two problems and the algorithms, $A_1=$DEMO^{SP2}, $A_2=$DEMO^{NS-II}, and $A_3=$NSGA-II

Ranking	DTLZ3			WFG8		
	A_1	A_2	A_3	A_1	A_2	A_3
1.00	3	4	0	4	0	0
2.00	1	0	0	0	3	3
3.00	0	0	4	0	1	1
Final	2.00	1.00	3.00	1.00	2.50	2.50

all quality indicators for each algorithm is further ranked with the fractional ranking scheme on each single problem separately.

To see what happens for a single problem when the hierarchical majority vote ensemble is utilized, Table 7.6 presents the process of ranking for the problems DTLZ3 and WFG8, separately.

By using the transformed DSC rankings (Table 7.4), the possible unique rankings for the DTLZ3 problem are 1.00, 2.00, and 3.00. Then for each algorithm, the number of times each unique ranking is obtained is counted. Using the Table 7.6 (the DTLZ3 problem), the algorithms are firstly compared concerning the best ranking (i.e., in our case 1.00). Using the best ranking, it follows that the $A_2=$DEMO^{NS-II} wins against the other two algorithms, so it is ranked as the best (i.e., ranking 1.00). Next, the comparison is done using the remaining algorithms, from which the $A_1,=$DEMO^{SP2} wins because it wins against the $A_3=$NSGA-II with regard to all four quality indicators, so it obtains the final ranking 2.00. The NSGA-II obtains the final ranking 3.00.

For the WFG8 problem, the DEMO^{SP2} wins against the other two algorithms according to four quality indicators and it obtains a ranking of 1.00. Then the other two remaining algorithms are compared. It seems that they are the same according to the best transformed DSC ranking (1.00), so the comparison continues with the next ranking (2.00). In this example, both algorithms are the same according to each unique ranking, so they obtain an average ranking, which means that there is no difference between their performance according to the set of used quality indicators.

To see what happens on a single problem when the data-driven ensemble is utilized, let us focus on the DLTZ5 problem. The matrix of the DSC rankings for each quality indicator and its normalization are presented in the upper part of Table 7.7. The transformed DSC rankings (Table 7.4) for the epsilon indicator on the DLTZ5 problem are 1.00, 1.00, and 1.00. This causes a problem in the normalization process, since the normalized rankings are indeterminate forms (i.e., 0/0) [51]. From here, it follows that the weight or the relative significance of this quality indicator cannot be calculated. From the other side, looking into the DSC rankings, it follows that they are all winners. Let us make an assumption that the weight w_3 could be calculated, so we need to calculate the part from the average preference index related to it. This part is the product of $w_3 P_3(A_{i_1}, A_{i_2})$, where $i_1, i_2 = 1, \ldots, m$ and $i_1 \neq i_2$. If we calculate

Table 7.7 Matrix with the DSC rankings for DLTZ5

Algorithm	Decision matrix				Normalized matrix			
	q_1	q_2	q_3	q_4	q_1	q_2	q_3	q_4
DEMO^{SP2}	2.00	1.00	1.00	1.00	0.00	1.00	0/0	1.00
DEMO^{NS-II}	2.00	1.00	1.00	3.00	0.00	1.00	0/0	0.00
NSGA-II	1.00	3.00	1.00	2.00	1.00	0.00	0/0	0.50
DEMO^{SP2}	2.00	1.00	/	1.00	0.00	1.00	/	1.00
DEMO^{NS-II}	2.00	1.00	/	3.00	0.00	1.00	/	0.00
NSGA-II	1.00	3.00	/	2.00	1.00	0.00	/	0.50

it, it follows that it will be zero, so it will not affect the average preference index used for calculating the positive and the negative preference flow. From this analysis, it follows that this quality indicator does not contribute to the ranking process, so it can be removed and the ranking can be performed using the remaining quality indicators (i.e., q_1, q_2, and q_4).

The matrix with the DSC rankings and its normalization when the epsilon indicator is removed are presented at the bottom part of Table 7.7. The weights obtained for the three quality indicators using the Shannon entropy weighted method are 0.57, 0.20, and 0.23, respectively.

The rankings obtained by the PROMETHEE II method together with the outranking flows that lead to these rankings are presented in Table 7.8, whereas in Table 7.9, the average preference indices that are used for calculating the positive and negative flows for the problem DLTZ5 are presented.

If we compare the three ensemble heuristics on the DTLZ5 problem, we get that the rankings obtained using the average ensemble and the hierarchical majority vote are the same and are 1.00, 2.50, and 2.50 (see Table 7.5). Using the hierarchical majority vote, the DEMO^{SP2} is ranked first because it wins in three out of four quality

Table 7.8 Outranking flows and PROMETHEE II rankings for DLTZ5

Algorithm	ϕ^+	ϕ^-	ϕ	Ranking
DEMO^{SP2}	0.11	0.10	0.01	2.00
DEMO^{NS-II}	0.03	0.17	-0.14	3.00
NSGA-II	0.23	0.10	0.13	1.00

Table 7.9 Average preference indices for DLTZ5

	$\pi(A_i, A_1)$	$\pi(A_i, A_2)$	$\pi(A_i, A_3)$
$\pi(A_1, A_j)$	0.00	**0.08**	**0.14**
$\pi(A_2, A_j)$	**0.00**	0.00	**0.06**
$\pi(A_3, A_j)$	**0.19**	**0.27**	0.00

indicators, while DEMO^{NS-II} and NSGA-II are ranked second (e.g., 2.5) because both are ranked first in the case of two quality indicators, then both are second in the case of one quality indicator and third in the case of one quality indicator. We need to point out here that all quality indicators are equally treated.

In the case of the data-driven ensemble, the obtained rankings are 2.00, 3.00, and 1.00. From Table 7.8, it is obvious that the NSGA-II has the highest positive flow. The question that arises here is why it is ranked first when DEMO^{SP2} has two wins. This happens because the importance of the quality indicators that are involved in the comparison is estimated from the data, which is obtained by the Shannon entropy weighted method. The quality indicators are ordered as q_1, q_4, and q_2, (e.g., hypervolume, generational distance, and r_2 indicator), starting from the most significant one to the least significant one. The average preference indices between A_1 and A_3 that are used for calculating the positive and negative flows are presented as follows:

$$\pi(A_1, A_3) = \frac{1}{3}[0.57 \cdot 0 + \mathbf{0.20 \cdot 1} + \mathbf{0.23 \cdot 1}] = 0.14$$

$$\pi(A_3, A_1) = \frac{1}{3}[\mathbf{0.57 \cdot 1} + 0.20 \cdot 0 + 0.23 \cdot 0] = 0.19. \tag{7.9}$$

Using the above-presented results, the average preference index between NSGA-II and DEMO^{SP2} is 0.19, which is a result of only one win regarding the quality indicator q_1. The opposite, the average preference index between DEMO^{SP2} and NSGA-II, is 0.14, which is smaller even though DEMO^{SP2} has two wins with regard to q_2 and q_4. This happens because q_1 is the most significant and its weight is much higher than the sum of the weights of q_2 and q_4.

After the analysis presented for a single problem, the rankings obtained by the three ensemble heuristics for the set of sixteen problems (i.e., presented in Table 7.4) can be analyzed with an appropriate omnibus statistical test to find the statistical result for a multiple-problem scenario. In all cases, the Friedman test is the appropriate one, and for all three heuristics it provides a p-value around 0.00. This shows that there is a statistical significance in the performance between the compared algorithms using the selected set of benchmark problems and a significance level of 0.05.

When the number of algorithms involved in the comparison increases, it is better to use multiple Wilcoxon tests, one for each pairwise comparison, and then combine the p-values to find the actual p-value for this scenario. More about this scenario, which is the multiple comparisons with a control algorithm, is presented in Sect. 6.1.2.3. The main difference is that instead of the DSC ranking scheme, the ranking schemes of the ensemble heuristics should be used to transform the data for each pairwise comparison.

7.3   Multi-objective Deep Statistical Comparison

The end results of benchmark studies of multi-objective optimization algorithms are affected by the selection of the quality indicators involved in the study. Even more, not only the selection, but each quality indicator, by transforming the high-dimensional into one-dimensional data, loses information covered in the high-dimensional space that can affect the end result of the benchmarking [31]. To go beyond this, we recently proposed a novel approach, known as *Multi-objective Deep Statistical Comparison (moDSC)* [30], which compares the distribution of the approximation sets (i.e., high-dimensional data). The moDSC is based on the idea used in Deep Statistical Comparison [37] for comparing single-objective optimization algorithms. The main difference is the dimensionality of the data that is involved in the comparison. The DSC ranking scheme (introduced in Sect. 5.2.1) works by comparing the distributions of one-dimensional data in the form of vectors (i.e., in our case, obtained solution values from single-objective optimization algorithms). The eDSC rankings scheme (introduced in Sect. 6.3) works with high-dimensional data in the form of a matrix (i.e., in our case, obtained solution locations in the search space from single-objective optimization algorithms). In the case of the moDSC, the result from each multi-objective optimization algorithm is a set of approximation sets, or an array of matrices. Compared with the eDSC approach, the moDSC approach has an additional dimension that should be taken into consideration when the performance assessment is done.

The main benefit of the moDSC ranking scheme is that it reduces information loss by transforming high-dimensional data into one-dimensional data. With this, it reduces the influence of the users' preference or the selection of a quality indicator on the end benchmarking conclusions. The quality indicator is applied only when the distributions of the high-dimensional data differ, indicating a statistically significant difference between approximation sets.

7.3.1   Multi-Objective Deep Statistical Comparison Ranking Scheme

Let m be the number of algorithms, k the number of multi-objective problem instances involved in the benchmarking process, n the number of runs performed by each algorithm on a single-problem instance, w the number of solutions that belong to the approximation set obtained within an algorithm run (i.e., it can be a different value for different runs of the same algorithm on the same problem instance), and d the number of objectives.

Let $X_{i,l,j}$ be a $w \times d$ matrix, where, $i = 1, \ldots, k, l = 1, \ldots, m$, and $j = 1, \ldots, n$. The matrix $X_{i,l,j}$ has the following form:

$$X_{i,l,j} = \begin{bmatrix} x_1 \\ x_2 \\ \vdots \\ x_w \end{bmatrix}, \tag{7.10}$$

where the rows of the matrix $X_{i,l,j}$ are d-dimensional vectors, which are the solutions that belong to the approximation set obtained for the jth run of the lth algorithm on the ith problem instance. Since the compared algorithms are stochastic in nature, for each algorithm we obtained a set of matrices $\{X_{i,l,1}, X_{i,l,2}, \cdots, X_{i,l,n}\}$ for each optimization problem instance. Each matrix from the set of matrices corresponds to one run of the lth algorithm obtained on the ith problem instance. We should also mention here that the approximation sets that are the results from n runs of the lth algorithm on the ith problem instance are independent and derive from the same probability distribution.

The moDSC follows the idea of the DSC ranking scheme with the difference in the dimensionality of the data that is used in the comparison. For this purpose, the moDSC ranking scheme is a mix from the eDSC ranking scheme (introduced in Sect. 6.3) and the Monte Carlo pDSC ranking scheme (introduced in Sect. 6.2.1.2). For it, all pairwise comparisons should be performed and the p-values should be organized in a matrix defined with Eq. 5.1. To compare distributions of high-dimensional data, a *multivariate $\mathcal{E}$-test* was proposed in [100].

For each pairwise comparison, two sets of matrices (i.e., approximation sets), $\{X_{i,l_1,1}, X_{i,l_1,2}, \cdots, X_{i,l_1,n}\}$ and $\{X_{i,l_2,1}, X_{i,l_2,2}, \cdots, X_{i,l_2,n}\}$, are involved in the comparison. Each set corresponds to the result that is obtained by a single algorithm. Using the information that all approximation sets obtained for n independent runs of an algorithm on the same problem instance are coming from the same distribution, we should compare two algorithms on the same problem instance by comparing the distributions of pairs of approximations sets, where each pair contains one approximation set from both algorithms. The question that arises here is which pair of approximation sets should be selected for comparing the distributions. To compare the distributions of pairs of approximation sets and to obtain more robust statistical results, the moDSC ranking scheme follows the idea used in the Monte Carlo pDSC ranking scheme. So, before each pairwise comparison is done, for each algorithm different orders of the obtained approximation sets are generated using their permutations. The number of such permutations per algorithm is $n!$. From here, it follows that there are $(n!)^2$ combinations of pairs of approximation sets that can be involved in each pairwise comparison. Each combination has one approximation set from the l_1th algorithm and one approximation set from the l_2th algorithm. Next, the moDSC rankings scheme randomly selects N different combinations for each pairwise comparison and multivariate $\mathcal{E}$-test is used to compare distributions for each combination. Further, to select one p-value that will be assigned to each pairwise comparison from the set of N p-values, the same idea already introduced for Monte Carlo pDSC ranking scheme is used, where a new random variable V is introduced to follow the information about how many null hypotheses are rejected out of N (see Sect. 6.2.1.2).

After obtaining the p-values for each pairwise comparison, the same steps as in the DSC ranking scheme are performed: the p-values are corrected, the transitivity property is checked, and the rankings are assigned.

The key difference between the DSC and the eDSC ranking schemes is in the selection of the representative values that will be involved in the ranking process. In the DSC ranking scheme, if the distributions of the obtained solution values from multiple runs are different, the ranking must be made using a mean value from the obtained solutions from the multiple runs. In the eDSC ranking scheme, when the distributions of the obtained solutions in the search space are different, the hypervolume covered by the distribution is used as a metric to rank the algorithms based on the user preference (wider or narrow distribution). However, in the moDSC this is not possible to be done, since we are comparing an array of matrices. For this purpose, in cases when distributions of the approximation sets are different, we can use a single-quality indicator or ensemble of quality indicators to rank the algorithms.

7.3.2 Sensitivity Analysis of the Multivariate $\mathcal{E}$-test

Here, the focus is on sensitivity analysis of the multivariate $\mathcal{E}$-test, since we used it for comparing distributions in high dimensions that are required by the moDSC ranking scheme.

To find a statistical significance between the distributions of two approximation sets, the multivariate $\mathcal{E}$-test should reject the null hypothesis. For this purpose, the test statistic used by the test should be greater than a critical value for a given significance level:

$$\mathcal{E}_{w_1,w_2} > c_\alpha. \tag{7.11}$$

Let us assume that $x_1,\ldots,x_{w_1}$ and $y_1,\ldots,y_{w_2}$ are solutions (i.e., vectors) that belong to two approximation sets in R^d, $d \geq 2$. Further, we can denote the approximation sets as X and Y. The multivariate $\mathcal{E}$-test test statistic between them is defined as

$$\mathcal{E}_{w_1,w_2} = \frac{w_1 w_2}{w_1 + w_2} \left(\frac{2}{w_1 w_2} \sum_{i=1}^{w_1} \sum_{m=1}^{w_2} ||x_i - y_m|| \right.$$
$$- \frac{1}{w_1^2} \sum_{i=1}^{w_1} \sum_{j=1}^{w_1} ||x_i - x_j||$$
$$\left. - \frac{1}{w_2^2} \sum_{l=1}^{w_2} \sum_{m=1}^{w_2} ||y_l - y_m|| \right). \tag{7.12}$$

The first double sum in the above equation provides the in between distance of the solutions from the approximation sets, the second double sum gives the within distance between the solutions from one approximation set (i.e., X), and the third double sum gives the within distance between the solutions from the other approximation set (i.e., Y).

This test statistic has a degenerate two-sample V-statistic and there exists a constant c_α satisfying

$$\lim_{w \to \infty} P\left(\frac{w_1 w_2}{w_1 + w_2} V_{w_1, w_2} > c_\alpha\right) = \alpha$$

in cases when the test will reject the null hypothesis [100].

To understand when the test will reject the null hypothesis, let us simplify the analysis where the distributions between one approximation set and its horizontal translation are investigated. In such a case, $w_1 = w_2 = w$. The translation is performed by moving each solution from the approximation set for a fixed step, s, in all objectives. To find the relation when distributions differ, we begin by analyzing the test statistic.

$$
\begin{aligned}
\mathcal{E}_{w,w} = \frac{w^2}{2w} \bigg(& \frac{2}{w^2} \sum_{i=1}^{w} \sum_{m=1}^{w} ||x_i - y_m|| \\
& - \frac{1}{w^2} \sum_{i=1}^{w} \sum_{j=1}^{w} ||x_i - x_j|| \\
& - \frac{1}{w^2} \sum_{l=1}^{w} \sum_{m=1}^{w} ||y_l - y_m|| \bigg).
\end{aligned}
\tag{7.13}
$$

The second approximation set is only a horizontal translation in all objectives from the first one, so it follows that both sets have the same within distance:

$$\sum_{i=1}^{w} \sum_{j=1}^{w} ||x_i - x_j|| = \sum_{l=1}^{w} \sum_{m=1}^{w} ||y_l - y_m||.
\tag{7.14}$$

Combining Eqs. 7.14, 7.13, and 7.11, it follows that

$$\sum_{i=1}^{w} \sum_{m=1}^{w} ||x_i - y_m|| > \sum_{i=1}^{w} \sum_{j=1}^{w} ||x_i - x_j|| + wc_\alpha.
\tag{7.15}$$

Let us denote the between and within distances as

$$B = \sum_{i=1}^{w} \sum_{m=1}^{w} ||x_i - y_m||,$$

$$W = \sum_{i=1}^{w} \sum_{j=1}^{w} ||x_i - x_j||. \tag{7.16}$$

The critical value, c_α, comes from a degenerate two-sample V-statistic. There is no statistical table from which the critical value can be obtained, however, there exist bootstrap methods for calculating the V-statistics [2]. To obtain the critical value of the two-sample degenerate V-statistic, we used the implementation of the multivariate $\mathcal{E}$-test, which is a permutation test. To find it, the R implementation of the multivariate $\mathcal{E}$-test returns the permutation replicates that are used to estimate it. For a large joint sample, the critical value is estimated as the $(1 - \alpha)$ quantile of the vector of replicates. This test is not parametric, and the critical value will vary for every possible X and Y. To get a good estimate of the critical value, we need to use a large number of replicates because the standard error of a quantile is high when the probability is small.

Figure 7.2 presents a geometrical interpretation of Eq. 7.15. The solutions space, resulting from the difference between the distributions of the solutions from the approximation set and its translation, is represented by the relation (red) of the between distance (B) and the critical value (c_α).

To check Eq. 7.15 in empirical settings, one approximation set obtained from the algorithm DEMO[SP2] on the DTLZ2 problem is analyzed, when the number of

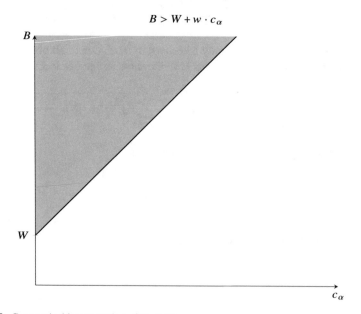

Fig. 7.2 Geometrical interpretation of Eq. 7.15

objectives is two. The approximation is obtained from the 30th run of the algorithm. Further, several translations were performed, where each solution is moved for some fixed step for all objectives. The set of fixed steps that was considered consists of the following steps: $1e$-09, $1e$-08, $1e$-07, $1e$-06, $1e$-05, $1e$-04, $1e$-03, $1e$-02, $1e$-01, $1e$+00, $1e$+01, $1e$+02, and $1e$+03. By experimental evaluation of the above-presented steps, we found the step value (i.e., $s = 1e$-01) where the statistical significance can be found. Further, the empirical evaluation was performed on the set of fixed steps in the range from $1e$-02 to $1e$+00.

Table 7.10 presents the results from the sensitivity analysis of the multivariate $\mathcal{E}$-test. For each translation of the approximation set, the step (s), the value of the test statistic $(\mathcal{E}_{w,w})$, the p-value, the critical value (c_α), the between distance (B), and the within distance (W) are presented. In addition, the condition represented by Eq. 7.15 is checked. In this example, the number of replicates used for calculating the critical value is $1e$+06.

The approximation set and all of the translations obtained for the steps presented in Table 7.10 are shown in Fig. 7.3. The front presented with circles corresponds to the approximation set. The next five blue fronts correspond to the translations obtained for the first five steps (Table 7.10), for which there is no statistical significance

Table 7.10 Sensitivity analysis of the multivariate $\mathcal{E}$-test

s	$\mathcal{E}_{100,100}$	p_{value}	c_α	B	W	Condition
0.01	0.06	1.00	1.31	4871.31	4865.21	✗
0.02	0.21	0.77	1.31	4885.78	4865.21	✗
0.03	0.42	0.42	1.31	4907.16	4865.21	✗
0.04	0.69	0.21	1.31	4934.53	4865.21	✗
0.05	1.02	0.10	1.31	4967.27	4865.21	✗
0.06	1.40	0.04	1.31	5004.88	4865.21	✓
0.07	1.82	0.02	1.31	5046.97	4865.21	✓
0.08	2.28	0.01	1.30	5093.21	4865.21	✓
0.09	2.78	0.00	1.30	5143.32	4865.21	✓
0.10	3.32	0.00	1.30	5197.03	4865.21	✓
0.20	10.28	0.00	1.31	5892.91	4865.21	✓
0.30	19.29	0.00	1.35	6793.93	4865.21	✓
0.40	29.62	0.00	1.45	7826.89	4865.21	✓
0.50	40.84	0.00	1.58	8949.46	4865.21	✓
0.60	52.70	0.00	1.74	10135.14	4865.21	✓
0.70	65.01	0.00	1.93	11366.56	4865.21	✓
0.80	77.67	0.00	2.14	12631.95	4865.21	✓
0.90	90.58	0.00	2.37	13923.08	4865.21	✓
1.00	103.69	0.00	2.60	15234.10	4865.21	✓

✗—Indicates that Eq. 7.15 is not satisfied
✓—Indicates that Eq. 7.15 is satisfied

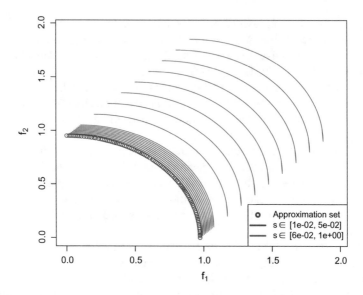

Fig. 7.3 Translations of one approximation set obtained from the algorithm DEMO^{SP2} on the DTLZ2 problem

between the distributions of the approximation set and its translations, while the last fourteen red fronts correspond to the translations, for which there is a statistical significance between the distributions of the solutions from the approximation set and its translation.

Equation 7.15 does not depend on the shape of the approximation set. Its evaluation of different shapes of fronts can be found in [30].

In a more general scenario, i.e., a comparison is made between two approximation sets, where $w_1 \neq w_2$, obtained from two different algorithms. In such cases, Eq. 7.15 becomes

$$\frac{2}{w_1 w_2} \sum_{i=1}^{w_1} \sum_{m=1}^{w_2} ||x_i - y_m|| >$$

$$\frac{1}{w_1^2} \sum_{i=1}^{w_1} \sum_{j=1}^{w_1} ||x_i - x_j|| + \frac{1}{w_2^2} \sum_{l=1}^{w_2} \sum_{m=1}^{w_2} ||y_l - y_m||$$

$$+ \frac{w_1 + w_2}{w_1 w_2} c_\alpha . \tag{7.17}$$

Equation 7.17 provides the information that the difference between the distributions of two approximation sets can be detected if the normalized between distance of the solutions is greater than the sum of the normalized within distances from both approximation sets plus a factor that contains the influence of the critical value.

7.3.3 Examples

To show how the moDSC ranking scheme works, let us assume an example, where a statistical comparison is performed between the algorithms $DEMO^{SP2}$, NSGA-II, and SPEA2 using the sixteen multi-objective optimization problems (i.e., introduced in Sect. 7.1.1.1). The number of objectives for all problems is set to 2. To explore the difference between the moDSC ranking scheme and the DSC ranking scheme (see Sect. 5.2.1) that works with single-quality-indicator data (i.e., in our experiment with the hypervolume), the rankings are calculated by both ranking schemes. The DSC ranking scheme compares the algorithms comparing the distributions of the hypervolume data, while the moDSC compares the high-dimensional distributions of the approximation sets and only if the distributions are statistically significant, the algorithms are compared using the DSC ranking scheme applied to the hypervolume data. For further illustration, let's denote both ranking schemes as DSC-SingleQI and moDSC-SingleQI.

Table 7.11 presents the algorithms' rankings obtained by both ranking schemes for all sixteen problems. Comparing the obtained rankings, it is obvious that there exist problems when the rankings differ. For example, moDSC-SingleQI shows that there is no statistical significance between the performance of the algorithms for the problems: DTLZ2, DTLZ5, WFG3, WFG4, WFG5, and WFG6. However, the DSC-SingleQI shows that there is a statistical significance between the compared

Table 7.11 Rankings for the algorithms $A_1=DEMO^{SP2}$, $A_2=NSGA-II$, and $A_3=SPEA2$

F	DSC-SingleQI			moDSC-SingleQI		
	A_1	A_2	A_3	A_1	A_2	A_3
DTLZ1	1.00	3.00	2.00	1.00	3.00	2.00
DTLZ2	1.00	3.00	2.00	2.00	2.00	2.00
DTLZ3	1.00	3.00	2.00	1.00	3.00	2.00
DTLZ4	1.00	3.00	2.00	1.00	3.00	2.00
DTLZ5	1.00	3.00	2.00	2.00	2.00	2.00
DTLZ6	1.00	3.00	2.00	1.00	3.00	2.00
DTLZ7	1.00	3.00	2.00	1.50	3.00	1.50
WFG1	1.00	2.50	2.50	1.00	2.50	2.50
WFG2	1.00	2.50	2.50	1.00	2.50	2.50
WFG3	1.00	3.00	2.00	2.00	2.00	2.00
WFG4	1.00	3.00	2.00	2.00	2.00	2.00
WFG5	1.00	3.00	2.00	2.00	2.00	2.00
WFG6	1.00	3.00	2.00	2.00	2.00	2.00
WFG7	1.00	3.00	2.00	1.00	3.00	2.00
WFG8	2.00	1.00	3.00	2.00	1.00	3.00
WFG9	2.00	2.00	2.00	2.00	2.00	2.00

Table 7.12 P-value matrix and its binarization for the WFG5 problem

	A_1	A_2	A_3
A_1	1.00	0.30	0.33
A_2	0.30	1.00	0.33
A_3	0.33	0.33	1.00
	A_1	A_2	A_3
A_1	1.00	1.00	1.00
A_2	1.00	1.00	1.00
A_3	1.00	1.00	1.00

algorithms and they obtained different rankings using the DSC ranking scheme. For the problems: DTLZ1, DTLZ3, DTLZ4, DTLZ6, WFG1, WFG2, WFG7, WFG8, and WFG9, the DSC-SingleQI provides the same result as the moDSC-SingleQI. For the problem DTLZ7, the DSC-SingleQI rankings are 1.00, 3.00, and 2.00 and the moDSC-SingleQI rankings are 1.50, 3.00, and 1.50. To understand why differences appear in the ranking process, three problems (i.e., WFG5, DTLZ7, and WFG8) are selected to be presented in more detail.

To calculate a p-value for each pairwise comparison that will be involved in the moDSC ranking scheme, 1000 combinations (i.e., pairs of approximation sets) were used. Further, for each combination the multivariate $\mathcal{E}$-test was used to calculate a p-value. To select a representative p-value from the set of 1000 p-values obtained for the same pairwise comparison, Eq. 6.4 was used with a prior level of significance $\alpha_p = 0.05$. After selecting a p-value for each pairwise comparison, they were further corrected using the Bonferroni correction. We should point out here that for each benchmark problem, all 30 runs from each algorithm are involved in the comparison. The sensitivity analysis of the moDSC ranking scheme when the number of runs is less than 30 is presented in [30].

Table 7.12 provides the corrected p-values for each pairwise comparison obtained for the WFG5 problem (left part) and the binarization of this matrix required to check the transitivity (right part).

Looking into the binarization, it follows that the transitivity of the p-value matrix is satisfied. However, here the set of algorithms is not divided into disjoint sets because all algorithms belong to the same set. This result indicates that all algorithms have the same distribution of the solutions in the approximation sets, so they are ranked as 2.00, 2.00, and 2.00, respectively.

To clarify this, Fig. 7.4 presents the empirical probability density functions for the solutions of one approximation set (i.e., a single run) obtained from each algorithm on the problem WFG5. Using the visualization, it follows that there is no difference between the distributions of solutions in the approximation set. Using this result, the algorithms are ranked the same, so there is no need to select a quality indicator (user preference) in order to distinguish which algorithm provides the best result.

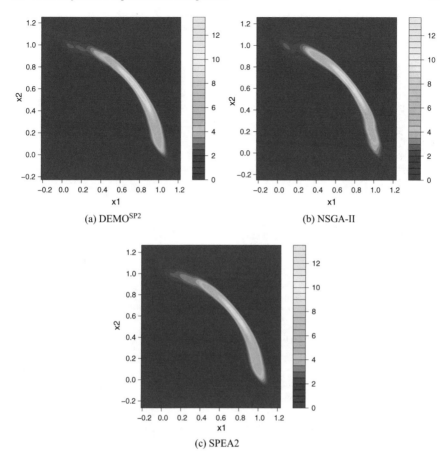

Fig. 7.4 Contour plots for empirical probability density functions for the solutions of one approximation set obtained on the problem WFG5

In the case when the DSC-SingleQI is used, the rankings are 1.00, 3.00, and 2.00. Here, the DSC ranking scheme compares the distribution of the hypervolume data obtained by each algorithm on the WFG5 problem. Figure 7.5 presents the empirical cumulative distributions of the hypervolume data for WFG5. Using the visualization, it is obvious that they differ. This result is further checked and confirmed using the two-sample AD test. From here, it follows that the algorithms are ranked according to the mean value of the hypervolume data. The algorithm with higher value of the hypervolume mean is ranked as better. This indicates that by transforming the high-dimensional data (i.e., approximation sets) to one-dimensional data (i.e., quality indicator) we can lose some information from the high-dimensional space that can further result in statistically significant differences in the one-dimensional space, even if they are not present in the high-dimensional space.

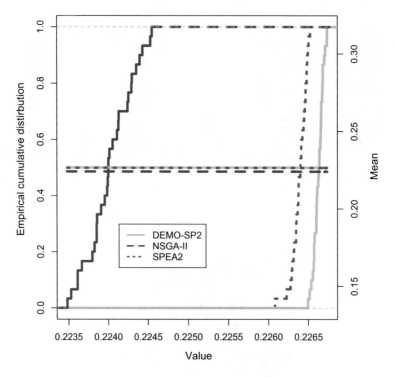

Fig. 7.5 Empirical cumulative distributions (step functions) and the mean values (horizontal lines) of the hypervolume data for WFG5

Table 7.13 P-value matrix and its binarization for the DTLZ7 problem

	A_1	A_2	A_3
A_1	1.00	0.01	0.32
A_2	0.01	1.00	0.01
A_3	0.32	0.01	1.00
	A_1	A_2	A_3
A_1	1.00	0.00	1.00
A_2	0.00	1.00	0.00
A_3	1.00	0.00	1.00

The p-value matrix and its binarization for the DTLZ7 problem are presented in Table 7.13. The transitivity of the p-value matrix is satisfied, so the set of algorithms is split into two disjoint sets {DEMO^{SP2}, SPEA2}, and {NSGA-II}. This result indicates that both algorithms DEMO^{SP2} and SPEA2 have the same distribution of the solutions in the approximation sets, which differs from the distribution provided by the algorithm NSGA-II. The result can be also seen in Fig. 7.6, where the empirical

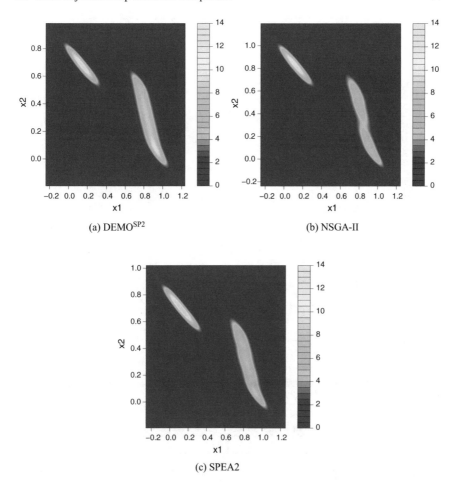

Fig. 7.6 Contour plots for empirical probability density functions for the solutions of one approximation set obtained on the problem DTLZ7

probability density functions for the solutions of one approximation set obtained from each algorithm on the problem DTLZ7 are presented.

The next step is to find the rankings that will be assigned to each of the disjoint sets. In our experiment (as already mentioned), we decided in such a situation to involve the DSC ranking scheme applied to the quality-indicator data. The rankings that are obtained by the DSC-SingleQI are 1.00, 3.00, and 2.00 (see Fig. 7.7). Because the first disjoint set contains DEMO^{SP2} and SPEA2, which are ranked as 1.00 and 2.00, respectively, on average (1.50) they are better than the NSGA-II, so the moDSC rankings are 1.50, 3.00, and 1.50.

Focusing on the WFG8 problem, the p-value matrix and its binarization are presented in Table 7.14. The transitivity of the p-value matrix is not satisfied, so all

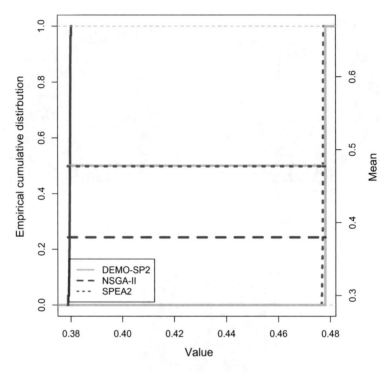

Fig. 7.7 Empirical cumulative distributions (step functions) and the mean values (horizontal lines) of the hypervolume data for DTLZ7

Table 7.14 P-value matrix and its binarization for the WFG8 problem

	A_1	A_2	A_3
A_1	1.00	0.01	0.00
A_2	0.01	1.00	0.00
A_3	0.00	0.00	1.00
	A_1	A_2	A_3
A_1	1.00	0.00	0.00
A_2	0.00	1.00	0.00
A_3	0.00	0.00	1.00

algorithms have different distributions of solutions in the approximation sets (see Fig. 7.8). Next, the algorithms obtain their rankings according to the DSC-SingleQI with hypervolume in this experiment. The rankings obtained are 2.00, 1.00, and 3.00 (see Fig. 7.9).

If we look at all sixteen problems, in seven out of them the loss of information is presented, due to the transformation of the high-dimensional data into one-dimensional data. Further, the rankings obtained by both ranking schemes (presented

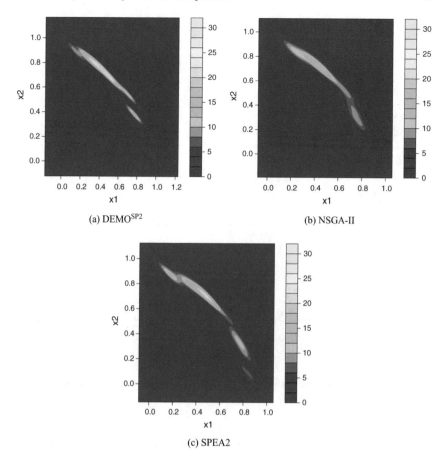

Fig. 7.8 Contour plots for empirical probability density functions for the solutions of one approximation set obtained on the problem WFG8

in Table 7.11) for all problems are analyzed with the Friedman test. In the case of the DSC-SingleQI, the p-value is 0.00, which indicates that there is a statistical significance between the performance of the algorithms on the set of sixteen problems. However, using the DSC-SingleQI, the p-value is 0.05, which indicates that there is no statistical significance between the performance of the algorithms. This result indicates that we should take great care when the high-dimensional data is transformed into one-dimensional data, since this can affect the statistical outcome of the study.

More information about the influence of the selection of quality indicators (e.g., epsilon, r_2, and generational distance) and more importantly reduction of influence using moDSC ranking scheme is presented in [30].

We would like to point here that in cases when the distributions of the approximation sets in the high-dimensional space are different, in the presented examples we

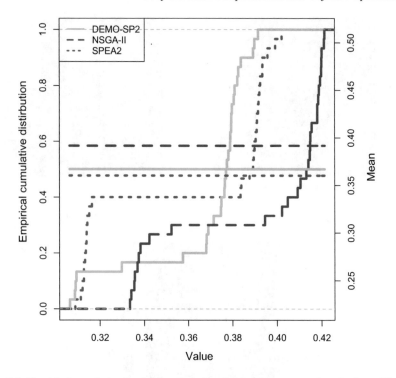

Fig. 7.9 Empirical cumulative distributions (step functions) and the mean values (horizontal lines) of the hypervolume data for WFG8

used the DSC ranking scheme with a single-quality indicator. However, in such cases users can also continue to use one of the three presented ensemble heuristics (i.e., average, hierarchical majority vote, and data-driven presented in Sects. 7.2.1, 7.2.2, and 7.2.3).

7.4 Summary

In this chapter,

- We introduced through examples how the Deep Statistical Comparison approach can be used for comparing the performance of multi-objective optimization algorithms using a single-quality indicator.
- We introduced three ensemble heuristics that combine the information obtained for a set of user-specified quality indicators in order to provide a more general conclusion in assessing the performance of multi-objective optimization algorithms. This approach reduces the bias in reporting results that are in favor of the winning algorithm.

- We introduced a multi-objective Deep Statistical Comparison approach that ranks the algorithms based on the distribution of the approximation sets. This approach reduces information loss when high-dimensional data is transformed into one-dimensional data. In addition, it reduces the influence of the users' preference in the selection of a quality indicator on the end benchmarking result.
- All examples presented in this chapter can be repeated and reproduced using code presented in Chap. 8.

Chapter 8
DSCTool—A Web-Service-Based e-Learning Tool

8.1 DSCTool

The DSCTool was developed and implemented to make all the required knowledge for making different performance evaluations accessible from a single place [41]. It guides the user from providing input data (i.e., optimization algorithm results) and selection of the desired comparison scenario to the final result of the comparison. The implementation provides natural progress for all the steps of the performance evaluation, so no extra knowledge is required from the user since it was designed as an e-learning tool. The DSCTool pipeline is shown in Fig. 8.1 with all the required details presented in the following sections.

8.2 REST Web Services

A web service is software that supports one or more open protocols and standards defined for exchanging data between applications or systems, and makes itself available over the Internet. Due to the use of open protocols and standards, the software can be written in an arbitrary programming language and run on various platforms. This makes the web service widely interoperable and decoupled from the implementations that use its functionalities. It allows access to the tool from simple REST clients (e.g., cURL [16]), custom implementation in various programming languages, to website accesses. The protocols most commonly used are the Simple Object Access Protocol (SOAP) and REpresentational State Transfer (REST). In DSCTool, the REST protocol was implemented because of its simplicity, flexibility, and lightweight. Since REST is based on HTTP, we can use basic HTTP methods, such as GET, PUT, POST, and DELETE. A REST web-service access point is defined by a Uniform Resource Identifier (URI), which is a string of characters that unambiguously

© The Author(s), under exclusive license to Springer Nature Switzerland AG 2022
T. Eftimov and P. Korošec, *Deep Statistical Comparison for Meta-heuristic Stochastic Optimization Algorithms*, Natural Computing Series,
https://doi.org/10.1007/978-3-030-96917-2_8

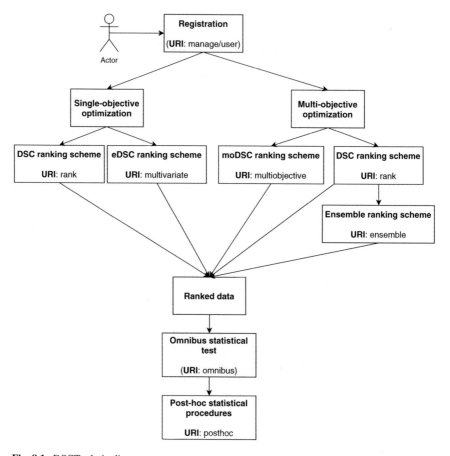

Fig. 8.1 DSCTool pipeline

identifies an individual resource. Data that is being transmitted can use different standard representations, such as XML or JSON.

The latest version (1.5, at the time of writing this book) of DSC web services is accessible from the following HTTPS URL https://ws.ijs.si/dsc/service/. For specifically accessing version 1.5, the following HTTPS URL https://ws.ijs.si/dsc/1.5/service/ should be used. All web services use JSON for input/output data representation. Data for all examples used in the book is available at Zenodo repository [29].

DSCTool consists of seven web services, each one (except registration) developed to cover different versions of the DSC ranking scheme:

- Registration service—used to register for all the web services. The output is information as to whether the registration succeeded, which grants the user free access to all the DSCTool web services.
- Ranking service—used to perform the DSC, sequential pDSC, or Monte Carlo pDSC ranking schemes with respect to appropriate input data. The output of this

web service is ranked data (i.e., the algorithms are ranked for each problem instance using the selected ranking scheme).

- Multivariate service—used to rank single-objective optimization algorithm results with respect to the distribution of the solutions in the search space. The ranking uses the eDSC ranking scheme. The output of this web service is again ranked data, like in the case with the ranking service.
- Multi-objective service—used to rank multi-objective optimization algorithm results obtained with respect to the distribution of approximation sets and the selected quality indicator. The ranking uses the moDSC ranking scheme. The output of this web service is again ranked data, like in the case with the ranking service.
- Ensemble service—used to rank multi-objective optimization algorithm results with respect to a set of performance measures (i.e., quality indicators). The ranking uses one of the three available ensemble heuristics: the average ensemble, the hierarchical majority vote ensemble, and the data-driven ensemble. The output of this web service is again ranked data, like in the case with the ranking service.
- Omnibus service—used to apply an appropriate omnibus statistical test using ranked data obtained by one of the above-presented web services. The output of this web service is a p-value and information as to whether the null hypothesis is rejected or not.
- Post-hoc service—used to perform multiple comparisons with a control algorithm in cases when the null hypothesis is rejected by the omnibus test. It provides information as to whether a statistical significance can be found between each compared algorithm and a control algorithm.

Next, more details for each web service are presented.

8.2.1 Registration Service

To prevent abuse of the web server, a mandatory registration is incorporated. Before using any of the DSC web services, a user must register using this service, accessed using "manage/user" URI. As input data, the user should provide their name and surname, affiliation, and email. In addition, the user must set a username and password. The output of this web service is the status information as to whether the registration process was successful. After the registration is completed, the user can perform all the DSC analyses provided by the web services. More details about the web service together with its input and output JSON are presented as follows:

URI manage/user
HTTP method POST
Input JSON

```
{
  "name": "Name Surname",
  "affiliation": "affiliation details",
```

```
     "email":  "name.surname@email.provider",
     "username":  "your_username",
     "password":  "your_password"
  }
```

Output JSON

```
  {
     "status":  "Done"|"Failed",
     "success":  boolean_value
  }
```

All passwords are encrypted before being stored in a database. To access any other service, a username and password using HTTP Basic Authentication are required.

8.2.2   Ranking Service

This service is used to perform the ranking by using the DSC ranking scheme (see Sect. 5.2.1), the sequential pDSC ranking scheme (see Sect. 6.2.1.1) or the Monte Carlo pDSC ranking scheme (see Sect. 6.2.1.2). It can be applied to the solution values in single-objective optimization or the calculated quality indicators in multi-objective optimization. The output is ranked data where all the involved algorithms are ranked on each individual problem using the selected ranking scheme. The ranked data can be used as input data for the omnibus service.

The input data consists of several parameters that must be provided by the user together with the data obtained by each algorithm that is to be compared.

More details about the web service are presented as follows:

URI   rank
HTTP method   POST
Input JSON

```
  {
     "epsilon":  double_value,
     "monte_carlo_iterations":  integer_value,
     "method":  {
       "name":  "KS"|"AD",
       "alpha":  double_value
     },
     "data":  [
       {
         "algorithm":  "algorithm name",
         "problems":  [
           {
             "name":  "problem name",
             "data":  [
               double_value,
                 ...
             ]
           },
```

```
        ...
      ]
    },
    ...
  ]
}
```

Output JSON

```
{
  "result": {
    "valid_methods": [
      "omnibus_test_name",
      ...
    ],
    "ranked_matrix": [
      {
        "problem": "problem name",
        "result": [
          {
            "algorithm": "algorithm name",
            "rank": double_value
          },
          ...
        ]
      },
      ...
    ],
    "number_algorithms": integer_value,
    "parametric_tests": boolean_value
  },
  "success": boolean_value
}
```

The *epsilon* key value defines the practical threshold used by the pDSC ranking schemes. If it is set to zero, the DSC ranking scheme will be applied. If it is greater than zero, the pDSC ranking scheme will be applied.

The *monte_carlo_iterations* key value defines the number of permutations needed for the Monte Carlo pDSC ranking scheme. This key is only used in the case when the *epsilon* key value is greater than zero (pDSC ranking scheme applied), otherwise the key is ignored. If the key is zero, the sequential pDSC ranking scheme is applied, otherwise the Monte Carlo pDSC is applied.

The *method* key defines the two-sample test that will be used to compare the distributions of one-dimensional data. There are two possibilities to choose from: the two-sample test Kolmogorov-Smirnov test (KS) or the two-samples Anderson-Darling test (AD).

The *alpha* key corresponds to the significance level used by the selected statistical test for comparing one-dimensional distributions.

The optimization results are described with the *data* key, where each algorithm's sample set is described with the algorithm's name and a list of problems containing a problem name and its solution values (single-objective optimization) or calculated quality indicators (multi-objective optimization).

In the case of a successful ranking (indicated by the *success* key) from the output JSON, a list of appropriate, supported omnibus tests (*valid_methods* key) is returned, together with the number of algorithms ranked (*number_algorithms* key), an indication of whether a parametric test is eligible (*parametric_tests* key), and the list of ranks (*ranked_matrix* key).

Each item in the ranked_matrix is defined by the name of the problem (*problem* key) and the rankings for each algorithm (*result* key) obtained on that problem. In cases when there is an insufficient number of problems to make an omnibus statistical test, the *valid_methods* key will contain the message "Not enough problems for a multiple-problem analysis!".

8.2.3   Multivariate Service

The multivariate service is used to compare single-objective optimization algorithms based on the distribution of the obtained solutions in the search space. It performs the eDSC ranking scheme defined in Sect. 6.3.

More details about the web service are presented as follows:

URI   multivariate
HTTP method   POST
Input JSON

```
{
  "method": {
    "desired_distribution": integer_value,
    "alpha": double_value
  },
  "data": [
    {
      "algorithm": "algorithm name",
      "problems": [
        {
          "name":"problem name",
          "data":[
            [
              double_value,
              ...
            ],
            ...
          ]
        },
        ...
      ]
    },
    ...
  ]
}
```

Output JSON   The output of this service is exactly the same as the rank service.

The *desired_distribution* key defines the preferred distribution of solutions, with 0 indicating a clustered distribution and 1 indicating a sparse distribution.

The *alpha* key defines the significance level used by the statistical test for comparing distributions.

The *data* key structure is similar to the one found in the rank service with the difference that the *data* key consists of an array of double values describing the solution locations and not its values.

The output of this service is exactly the same as the rank service, from which the ranked matrix should be further used as input data for the omnibus service.

8.2.4 Ensemble Service

The rank service defined in Sect. 8.2.2 can be used to compare multi-objective optimization algorithms using a single-quality-indicator data. However, when a more general conclusion should be obtained, the ranking can be made using a set of quality indicators. For this purpose, the ensemble service should be used, where the user needs to select one of the three predefined ensemble heuristics (see Sect. 7.2): the average, the hierarchical majority vote, or the data-driven heuristic.

More details about the service are presented as follows:

URI   ensemble
HTTP method   POST
Input JSON

```
{
  "method": "average"|"hierarchical"|"data-driven",
  "problems": [
    "problem_name",
    ...
  ],
  "indicators": [
    {
      "indicator": "indicator name",
      "data": [
        {
          "algorithm": "algorithm name",
          "problems": [
            double_value,
            ...
          ]
        },
        ...
      ]
    },
    ...
  ]
}
```

Output JSON   The output of this service is exactly the same as the rank service.

The *method* key defines the ensemble heuristic (average (Sect. 7.2.1), hierarchical majority vote (Sect. 7.2.2), or data-driven (Sect. 7.2.3)) that should be performed.

The *problems* key represents a list of problems on which algorithms were run.

The *ensembles* key represents a list of ensemble items, where each item is represented by an indicator name, and the list of rankings obtained for each algorithm on each problem. The rankings involved for each algorithm on each problem are the DSC rankings obtained when the algorithms are compared separately for each quality indicator by the DSC ranking scheme. This means that for each quality indicator, the algorithms should first be compared using the rank service, and the results obtained in the ranked matrices (i.e., one ranked matrix per quality indicator) should be used as the input data for the input JSON required for the ensemble web service.

The output of this service is exactly the same as the rank service, from which the ranked matrix should be further used as the input data for the omnibus service.

8.2.5   Multi-objective Service

It is well known that the selection of quality indicators can have a large influence on the end result of the statistical comparison. To limit the influence of the selection of quality indicators and not to lose information when high-dimensional approximation sets are transformed into one-dimensional quality-indicator data, the multi-objective service should be used. It performs the moDSC ranking scheme presented in Sect. 7.3.1. This web service is a really time-consuming service and can take more than 10 min to process, so please set the client's timeout appropriately.

More details about the web service are presented as follows:

URI   multiobjective
HTTP method   POST
Input JSON

```
{
  "method": {
    "alpha": double_value
  },
  "data": [
    {
      "algorithm": "algorithm name",
      "problems": [
        {
          "name": "problem name",
          "approximationSets": [
            {
              "preferenceData": double_value,
              "data":[
                [
                  double_value,
                  ...
```

```
                    ],
                      ...
                  ]
                },
                  ...
              ]
            },
              ...
          ]
        },
          ...
      ]
    }
```

Output JSON   The output of this service is exactly the same as the rank service.

The *alpha* key defines the significance level used by the statistical test for comparing high-dimensional distributions.

The *data* key structure consists of the key values *approximationSets* and *preferenceData*. The *approximationSets* structure is the same as the one found in the multivariate service with the difference that the array of double values describes one approximation set (i.e., the objectives for each solution in the approximation set). The *preferenceData* key is added to the structure and consists of the calculated quality-indicator value from the approximation set that should be used for ranking when the high-dimensional distributions are statistically significant.

The output of this service is exactly the same as the rank service, from which the ranked matrix should be further used as input data for the omnibus service.

8.2.6 Omnibus Service

After ranking the algorithms on a set of problem instances using an appropriate DSC ranking scheme, the next step is to analyze the obtained result (i.e., the DSC rankings) using an appropriate omnibus statistical test. So, no matter which web service was performed to obtain the rankings (either from the rank, multivariate, ensemble, or multi-objective service depending on the scenario), they need to be further analyzed using the omnibus service.

More details about this web service are presented as follows:

URI   omnibus
HTTP method   POST
Input JSON

```
{
  "method": {
    "name": "omnibus test name",
    "alpha": double_value
  },
```

```
"ranked_matrix": [
  {
    "problem": "problem name",
    "result": [
      {
        "algorithm": "algorithm name",
        "rank": double_value
      },
      ...
    ]
  },
  ...
],
"number_algorithms": integer_value,
"parametric_tests": boolean_value
}
```

Output JSON

```
{
  "result": {
    "message": "information if hypothesis is rejected or not",
    "p_value": double_value,
    "t": double_value,
    "method": {
      "name": "omnibus test name",
      "alpha": double_value
    },
    "algorithm_means": [
      {
        "algorithm": "algorithm name",
        "mean": double_value
      },
      ...
    ]
  },
  "success": boolean_value
}
```

The input JSON consists of *ranked_matrix* key value that can be taken from the ranking web service that is utilized to obtain the rankings.

In addition, the *omnibus test name* key value should be set, which is the name of the appropriate omnibus statistical test that should be utilized together with the *alpha* key value, which is the statistical significance. The name of the statistical test should be one of the choices that are recommended by any ranking service, since all ranking services check the required conditions for the safe use of the parametric test (i.e., the normality of the data and the homoscedasticity of variances) and provide a list of appropriate ones (see Sect. 4.5). All the ranking services are implemented as e-learning services.

The result of the omnibus service consists of the *message* key, which describes whether the null hypothesis is rejected or not, the *p_value* key indicating the obtained p-value, the *t* key representing the test statistic value (if appropriate, otherwise it is 0), the *name* of the omnibus statistical test that was applied together with its statistical significance (*alpha*), and a list of pairs that consist of the algorithm's name and the mean of the rankings calculated across all the benchmark problems.

We should point out here that if the "Wilcoxon-signed-rank-less" statistical test is selected, the p-value represents the result of a pairwise comparison of the algorithm with a smaller mean value versus the algorithm with a larger value. If the obtained p-value is smaller than the alpha value then the algorithm with the smaller mean significantly outperforms the other algorithm, otherwise there is no significance between them. This scenario is used in multiple comparisons with a control algorithm (see Sect. 6.1.2.3).

8.2.7   Post-Hoc Service

In cases when the omnibus service rejects the null hypothesis, we need to identify whether the selected algorithm performs better than the others using the post-hoc service.

More details about the post-hoc service are presented as follows:

URI   posthoc
HTTP method   POST
Input JSON

```
{
    "algorithm_means": [
      {
        "algorithm": "algorithm name",
        "mean": double_value
      },
      ...
    ],
    "k": integer_value,
    "n": integer_value,
    "base_algorithm": "control algorithm name",
    "method": {
      "name": "omnibus test name",
      "alpha": double_value
    }
}
```

Output JSON

```
{
    "result": [
      {
        "name": "ZValue",
        "algorithms": [
          {
```

```
            "algorithm": "algorithm name",
            "value": double_value
          },
          ...
        ]
      }, {
        "name": "UnadjustedPValue",
        "algorithms": [
          {
            "algorithm": "algorithm name",
            "value": double_value
          },
          ...
        ]
      }, {
        "name": "Holm",
        "algorithms": [
          {
            "algorithm": "algorithm name",
            "value": double_value
          },
          ...
        ]
      }, {
        "name": "Hochberg",
        "algorithms": [
          {
            "algorithm": "algorithm name",
            "value": double_value
          },
          ...
        ]
      }
    ],
    "success": boolean_value
}
```

The input JSON consists of *algorithm_means* key returned by the omnibus service.

The *algorithm_means* and *method* keys values can be taken from the output of the omnibus service.

The *k* key represents the number of compared algorithms.

The *n* key represents the number of problems.

The *base_algorithm* key defines the control algorithm (one of the algorithms defined inside the *algorithm_means* key).

The result of the post-hoc service consists of a list of four results, where a Z-value (ZValue) is the result of a selected statistic, an unadjusted p-value (UnadjustedP-Value) is calculated from the Z-value, and two adjusted p-values according to the Holm and Hochberg procedures are presented for each of the algorithms with respect to the selected control algorithm.

8.3 Clients

Due to this being a web-service-based implementation, there are many different options relating to how the DSCTool can be accessed. We can differentiate between two main options:

- General REST API clients.
- Specific API implementations in some programming language.

8.3.1 REST API Clients

There are many REST clients that can be used and are available on the Internet. Some of them provide a command-line interface (e.g., cURL [16]), while the others add GUI support (e.g., CocoaRestClient [78]), which makes them more user friendly. From the perspective of functionality, it does not matter which one is used. Both above-mentioned options require a lot of manual settings, since all of the DSCTool web services require data pre-processing to generate the appropriate input JSON file. This is acceptable for small sample data sets and only a few statistical comparisons that should be performed. In cases when many statistical comparisons should be performed, it is almost necessary to use some specific API implementation in a programming language of the user's choice. This is not a difficult task, since practically all popular programming languages already have libraries with support for REST API calls. This makes the implementation of clients for the DSCTool web services an easy task.

8.3.2 API Implementations in R

With specific API implementations, we are talking about clients being implemented specifically for the purpose of accessing the DSCTool web services. To make this possible, we present a DSCTool client written in the R programming language. However, since we can deal with a large amount of experimental data, there are already-implemented help R functions that transform the obtained optimization results into the required JSON inputs, depending on which web service will be used. To do this, the user should store the results in a predefined file format (e.g., in this case, a comma-separated file). The R client implementation together with the help functions and the predefined file format are available at Zenodo repository [29].

There are seven R functions implemented to return the required input JSON file. The R functions are as follows:

- Registration_client(postfield)—for performing a registration of a new user. It provides access to the registration service (see Sect. 8.2.1).

- DSC_client(username, password, postfield)—for performing the DSC, the sequential pDSC, or the Monte Carlo pDSC ranking scheme. It provides access to the rank service (see Sect. 8.2.2).
- eDSC_client(username, password, postfield)—for performing the eDSC ranking scheme. It provides access to the multivariate service (see Sect. 8.2.3).
- Ensemble_client(username, password, postfield)—for performing an average, a hierarchical majority vote, or a data-driven ensemble ranking using the results from multi-objective optimization. It provides access to the ensemble service (see Sect. 8.2.4).
- moDSC_client(username, password, postfield)—for performing the moDSC ranking scheme. It provides access to the multi-objective service (see Sect. 8.2.5).
- Omnibus_client(username, password, postfield)—for performing an omnibus statistical test. It provides access to the omnibus service (see Sect. 8.2.6).
- Posthoc_client(username, password, postfield)—for making post-hoc testing when the omnibus service rejects the null hypothesis. It provides access to the post-hoc service (see Sect. 8.2.7).

All the R functions except the registration function have three parameters: the username, the password, and the postfield. The username and the password are the authentications parameters that are specified by the user during their registration, while the postfield is the required JSON for each of the services. The registration client has only one parameter, which is the required input JSON.

8.4 Examples

In this section, we show how the examples used in Chaps. 6 and 7 can be performed using the DSCTool web services. First, examples for single-objective optimization will be presented, followed by examples presented for multi-objective optimization. All examples will be presented using cURL as a REST API client and functions implemented in the R programming language. For all the examples, it is expected that the user is already registered and has the valid username and password.

Example 1: Testing for statistical significance in single-objective optimization (DSC ranking scheme)

This example was introduced and explained in more detail in Sect. 6.1.2.2 (Table 6.2). Three single-objective optimization algorithms, GP5-CMAES, Sifeg, and BSif, are compared using benchmark problems defined in BBOB 2015. The dimension of the problems is set to 10. To test whether there is a statistical significance between their performances, the DSC ranking scheme is utilized with the two-sample AD test for comparing one-dimensional distributions with a statistical significance level of 0.05. To perform the comparison, the obtained solution values are organized in a JSON file required for the rank service (i.e., "dsc_ranking_in.json" file from the repository). To perform the ranking, we should run the following commands:

cURL:

```
curl --user username
  -H "Content-Type: application/json"
  -H "Accept: application/json"
  -X POST
  -d @dsc_ranking_in.json
  https://ws.ijs.si/dsc/service/rank
```

R:

```
postfield<-fromJSON("dsc_ranking_in.json")
result<-DSC_client(username, password, postfield)
```

The result is presented in the JSON file "dsc_ranking_out.json" available at the repository. Next, the ranked matrix that is obtained as a result is used to create the JSON file required for the omnibus service. In addition, the statistical significance level (i.e., alpha), the name of the omnibus statistical test that should be utilized (i.e., name), and a Boolean variable (i.e., parametric_test) that indicates whether the required conditions for the safe use of the parametric tests are satisfied should be set. The number of algorithms can be indirectly taken from the ranked matrix. All these parameters should be set and used to create the input JSON file required for the omnibus service "omnibus_in.json". Next, the omnibus service should be run using the following commands:

cURL:

```
curl --user username
  -H "Content-Type: application/json"
  -H "Accept: application/json"
  -X POST
  -d @omnibus_in.json
  https://ws.ijs.si/dsc/service/omnibus
```

R:

```
postfield<-fromJSON("omnibus_in.json")
result<-Omnibus_client(username, password, postfield)
```

The result is presented in the JSON file "omnibus_out.json" available at the repository. Next, if the obtained p-value is smaller than the significance level set, we should continue with a post-hoc test or perform multiple comparisons with a control algorithm. For this purpose, we should create the input JSON file required for the post-hoc service (i.e., "posthoc_in.json") available at our repository. Here, the result from the omnibus service should be reused from where the mean rankings for each algorithm should be taken together with the statistical test that was used in the omnibus test. Next, we should set the name of the control algorithm (i.e., base_algorithm), the number of algorithms involved in the comparison (i.e., k), and the number of benchmark problems (i.e., n). In our example, we have set GP1-CMAES as a control algorithm. To run the post-hoc service, we should perform the following commands.

cURL:

```
curl --user username
  -H "Content-Type: application/json"
  -H "Accept: application/json"
  -X POST
  -d @posthoc_in.json
  https://ws.ijs.si/dsc/service/posthoc
```

R:

```
postfield<-fromJSON("posthoc_in.json")
result<-Posthoc_client(username, password, postfield)
```

The result is presented in the JSON file "posthoc_out.json" available at the repository.

Example 2: Ranking single-objective optimization algorithms with regard to practical significance (pDSC ranking scheme)

This example was already introduced and explained in more detail in Sect. 6.2.2 (Table 6.8). Three single-objective optimization algorithms, RF1-CMAES, GP1-CMAES, and Srr, are compared using a set of benchmark problems defined in BBOB 2015, when the practical threshold is set to 10^{-1}. The dimension of the problems is set to 10. To rank the algorithms with regard to the practical threshold, we should first create the input JSON file required for the pDSC ranking scheme. In our case, the input file is available at the repository as "practical_ranking_in.json". The number of Monte Carlo permutations is set to zero, so the sequential pDSC ranking scheme will be applied. The practical threshold is set to 0.1. To perform the ranking, we should use the following commands:

cURL:

```
curl --user username
  -H "Content-Type: application/json"
  -H "Accept: application/json"
```

```
-X POST
-d @practical_ranking_in.json
https://ws.ijs.si/dsc/service/rank
```

R:

```
postfield<-fromJSON("practical_ranking_in.json")
result<-DSC_client(username, password, postfield)
```

The result is presented in the JSON file "practical_ranking_out.json" available at the repository. It consists of the rankings obtained for each algorithm on each benchmark problem (see Table 6.8). If we want to find whether there is a practical significance of the obtained results, the omnibus service and the posthoc service should be executed. We will not go into more detail here, since this was already described in Example 1, so the same commands and steps should be followed.

Example 3: Ranking single-objective optimization algorithms with regard to the distributions of the solutions in the search space (eDSC ranking scheme)

This example was already introduced and explained in more detail in Sect. 6.3.2.2 (Table 6.12). It consists of a comparison of three algorithms Cauchy-EDA, MCS, and iAMALGAM from the BBOB 2009 competition. The algorithms are compared using the set of benchmark problems when the dimension is set to 2. The ranking is performed for the distribution of the solutions in the search space (i.e., applying eDSC ranking scheme). The input JSON file for this example can be accessed at the repository as "multivariate_ranking_in.json". To apply the eDSC ranking scheme, we should run the following commands:

cURL:

```
curl --user username
  -H "Content-Type: application/json"
  -H "Accept: application/json"
  -X POST
  -d @multivariate_ranking_in.json
  https://ws.ijs.si/dsc/service/multivariate
```

R:

```
postfield<-fromJSON("multivariate_ranking_in.json")
result<-eDSC_client(username, password, postfield)
```

The result is presented in the JSON file "multivariate_ranking_out.json" available at the repository.

Example 4: Ranking multi-objective optimization algorithms with regard to single-quality-indicator data (DSC ranking scheme)

This example was already introduced and explained in more detail in Sect. 7.1.1.2 (Table 7.1). In this example, three multi-objective optimization algorithms, DEMO^{SP2}, DEMO^{NS-II}, and NSGA-II, are compared using a set of benchmark problems with regard to the hypervolume, epsilon, r_2, and the generational instance quality indicator. The number of objectives for each benchmark problem is set to 4. For this purpose, the rank service is used for each quality indicator separately. The input JSON files are available as "hypervolume_ranking_in.json", "epsilon_ranking_in.json", "r2_ranking_in.json", and "g_distance_ranking_in.json".

To rank them with regard to a single-quality indicator, the following commands should be used:

cURL:

```
curl --user username
   -H "Content-Type: application/json"
   -H "Accept: application/json"
   -X POST
   -d @hypervolume_ranking_in.json
   https://ws.ijs.si/dsc/service/rank
```

```
curl --user username
   -H "Content-Type: application/json"
   -H "Accept: application/json"
   -X POST
   -d @epsilon_ranking_in.json
   https://ws.ijs.si/dsc/service/rank
```

```
curl --user username
   -H "Content-Type: application/json"
   -H "Accept: application/json"
   -X POST
   -d @r2_ranking_in.json
   https://ws.ijs.si/dsc/service/rank
```

```
curl --user username
  -H "Content-Type: application/json"
  -H "Accept: application/json"
  -X POST
  -d @g_distance_ranking_in.json
  https://ws.ijs.si/dsc/service/rank
```

R:

```
pf_hyper<-fromJSON("hypervolume_ranking_in.json")
result_hyper<-DSC_client(username, password, pf_hyper)


pf_eps<-fromJSON("epsilon_ranking_in.json")
result_eps<-DSC_client(username, password, pf_eps)


pf_r2<-fromJSON("r2_ranking_in.json")
result_r2<-DSC_client(username, password, pf_r2)


pf_gd<-fromJSON("g_distance_ranking_in.json")
result_gd<-DSC_client(username, password, pf_gd)
```

The results are presented in JSON files "hypervolume_ranking_out.json", "epsilon_ranking_out.json", "r2_ranking_out.json", and "g_distance_ranking_out.json", available at the repository.

Example 5: Ranking multi-objective optimization algorithms with regard to set of quality indicators (average, hierarchical majority vote, and data-driven ensemble)

This example was already introduced and explained in more detail in Sect. 7.2.4 (Table 7.5). In this example, three multi-objective optimization algorithms, $DEMO^{SP2}$, $DEMO^{NS-II}$, and NSGA-II, are compared using a set of benchmark problems with regard to a set of quality indicators: hypervolume, epsilon, r_2, and generational instance. The difference with the previous example is that here we want to have one conclusion using all quality indicators, and not comparing the algorithms separately per each quality indicator. The number of objectives for each benchmark problem is set to 4. For this purpose, the ensemble service is used for each quality indicator separately. The input JSON files are available as "average_ranking_in.json", "hierarchical_ranking_in.json", and "ddriven_in.json". To create these JSON files, we need to reuse the results that were already obtained with the previous example ("hyper-volume_ranking_out.json", "epsilon_ranking_out.json", "r2_ranking_out.json", and "g_distance_ranking_out.json"). These files consist of the DSC rankings obtained for each algorithm on each benchmark problem for each quality indicator. These rankings are further combined in ensemble heuristics. To rank them with regard to a single-quality indicator, the following commands should be used:

cURL:

```
curl --user username
  -H "Content-Type: application/json"
  -H "Accept: application/json"
  -X POST
  -d @average_ranking_in.json
  https://ws.ijs.si/dsc/service/ensemble


curl --user username
  -H "Content-Type: application/json"
  -H "Accept: application/json"
  -X POST
  -d @hierachical_ranking_in.json
  https://ws.ijs.si/dsc/service/ensemble


curl --user username
  -H "Content-Type: application/json"
  -H "Accept: application/json"
  -X POST
  -d @ddriven_ranking_in.json
  https://ws.ijs.si/dsc/service/ensemble
```

R:

```
pf_avg<-fromJSON("average_ranking_in.json")
result_avg<-DSC_client(username, password, pf_avg)


pf_hierach<-fromJSON("hierachical_ranking_in.json")
result_hierach<-DSC_client(username, password, pf_hierach)


pf_ddriven<-fromJSON("ddriven_ranking_in.json")
result_ddriven<-DSC_client(username, password, pf_ddriven)
```

The results are presented in JSON files "average_ranking_out.json", "hierarchical_ranking_out.json", and "ddriven_ranking_out.json", available at the repository.

Example 6: Ranking multi-objective optimization algorithms with regard to the high-dimensional distribution of the approximation sets (moDSC ranking scheme)

This example was already introduced and explained in more detail in Sect. 7.3.3 (Table 7.11). In this example, three multi-objective optimization algorithms, DEMO^{SP2}, NSGA-II, and SPEA2 use a set of multi-objective optimization problems (i.e., introduced in Sect. 7.1.1.1). The number of objectives for all the problems is set to 2. Here, the moDSC is utilized, which is used to compare the algorithms based on the high-dimensional distribution of the obtained approximation sets. In cases when the distributions of the approximation sets obtained by different algorithms differ, then the DSC ranking with a single-quality indicator is used to select which algorithm is better. In our example, this is done using hypervolume. The input JSON file is available as "moDSC_ranking_in.json". To rank them using the moDSC ranking scheme, the following commands should be used:

cURL:

```
curl --user username
   -H "Content-Type: application/json"
   -H "Accept: application/json"
   -X POST
   -d @moDSC_ranking_in.json
   https://ws.ijs.si/dsc/service/multiobjective
```

R:

```
pf<-fromJSON("moDSC_ranking_in.json")
result<-DSC_client(username, password, pf)
```

The result is presented in the JSON file "moDSC_ranking_out.json", available at the repository.

8.5   Summary

In this chapter,

- We introduced the DSCTool that consists of all the presented Deep Statistical Comparison approaches for making a performance assessment of meta-heuristic stochastic optimization algorithms.
- We presented a web-service implementation of each Deep Statistical Comparison approach together with an explanation of the input and the output data required for using it.
- We showed examples of the performance assessment of meta-heuristic stochastic optimization algorithms (single- and multi-objective) using cURL, a REST API client, and functions implemented in the R programming language.
- All the examples presented in this chapter are explained in more detail in Chaps. 6 and 7.

References

1. Alridha, A., Salman, A.M., Al-Jilawi, A.S.: The applications of np-hardness optimizations problem. J. Phys.: Conf. Ser. **1818**, 012179 (IOP Publishing) (2021)
2. Arcones, M.A., Gine, E.: On the bootstrap of u and v statistics. In: The Annals of Statistics, pp. 655–674 (1992)
3. Atamna, A.: Benchmarking ipop-cma-es-tpa and ipop-cma-es-msr on the bbob noiseless testbed. In: Proceedings of the Companion Publication of the 2015 on Genetic and Evolutionary Computation Conference, pp. 1135–1142. ACM (2015)
4. Bajer, L., Pitra, Z., Holeňa, M.: Benchmarking gaussian processes and random forests surrogate models on the bbob noiseless testbed. In: Proceedings of the Companion Publication of the 2015 on Genetic and Evolutionary Computation Conference, pp. 1143–1150. ACM (2015)
5. Bartz-Beielstein, T., Doerr, C., Berg, D.V.D., Bossek, J., Chandrasekaran, S., Eftimov, T., Fischbach, A., Kerschke, P., La Cava, W., Lopez-Ibanez, M., et al.: Benchmarking in optimization: best practice and open issues. arXiv preprint arXiv:2007.03488 (2020)
6. Bates, D., Maechler, M.: Matrix: Sparse and Dense Matrix Classes and Methods (2017). URL https://CRAN.R-project.org/package=Matrix. R package version 1.2-8
7. Berger, V.W., Zhou, Y.: Kolmogorov-smirnov test: overview. In: Statistics Reference Online, Wiley Statsref (2014)
8. Bergmann, B., Hommel, G.: Improvements of general multiple test procedures for redundant systems of hypotheses. In: Multiple Hypothesenprüfung/Multiple Hypotheses Testing, pp. 100–115. Springer (1988)
9. Black Box Optimization Competition, B.: Black-box benchmarking 2015. http://coco.gforge.inria.fr/doku.php?id=bbob-2015. Accessed 01 Feb. 2016
10. Blum, C., Roli, A.: Metaheuristics in combinatorial optimization: Overview and conceptual comparison. ACM Comput. Surv. **35**(3), 268–308 (2003)
11. Boroushaki, S.: Entropy-based weights for multicriteria spatial decision-making. Yearb. Assoc. Pacific Coast Geogr. **79**, 168–187 (2017)
12. Brans, J.P., Vincke, P.: Note-A preference ranking organisation method: (the promethee method for multiple criteria decision-making). Manag. sci. **31**(6), 647–656 (1985)

© The Editor(s) (if applicable) and The Author(s), under exclusive license to Springer Nature Switzerland AG 2022

T. Eftimov and P. Korošec, *Deep Statistical Comparison for Meta-Heuristic Stochastic Optimization Algorithms*, Natural Computing Series,
https://doi.org/10.1007/978-3-030-96917-2

13. Brockhoff, D., Bischl, B., Wagner, T.: The impact of initial designs on the performance of matsumoto on the noiseless bbob-2015 testbed: a preliminary study. In: Proceedings of the Companion Publication of the 2015 on Genetic and Evolutionary Computation Conference, pp. 1159–1166. ACM (2015)

14. Cook, W.J., Cunningham, W.H., Pulleyblank, W.R., Schrijver, A.: Combinatorial Optimization. Wiley, New York (1997)

15. Cressie, N., Whitford, H.: How to use the two sample t-test. Biom. J. **28**(2), 131–148 (1986)

16. Curl: Curl: command line tool and library for transferring data with urls. https://curl.se/. Accessed 20 Oct. 2021

17. Dantzig, G.B., Cottle, R.W.: Positive (semi-) definite matrices and mathematical programming. Tech. rep, DTIC Document (1963)

18. Deb, K., Sindhya, K., Hakanen, J.: Multi-objective optimization. In: Decision Sciences: theory and Practice, pp. 145–184. CRC Press (2016)

19. Deb, K., Thiele, L., Laumanns, M., Zitzler, E.: Scalable Test Problems for Evolutionary Multiobjective Optimization. Springer (2005)

20. Demšar, J.: Statistical comparisons of classifiers over multiple data sets. J. Mach. Learn. Res. **7**, 1–30 (2006)

21. Derrac, J., García, S., Molina, D., Herrera, F.: A practical tutorial on the use of nonparametric statistical tests as a methodology for comparing evolutionary and swarm intelligence algorithms. Swarm Evolut. Comput. **1**(1), 3–18 (2011)

22. Dunn, O.J.: Multiple comparisons among means. J. Am. Stat. Assoc. **56**(293), 52–64 (1961)

23. Edelman, A., Rao, N.R.: Random matrix theory. Acta Num. **14**, 233–297 (2005)

24. Eftimov, T., Korošec, P.: The impact of statistics for benchmarking in evolutionary computation research. In: Proceedings of the Genetic and Evolutionary Computation Conference Companion, pp. 1329–1336 (2018)

25. Eftimov, T., Korošec, P.: Identifying practical significance through statistical comparison of meta-heuristic stochastic optimization algorithms. Appl. Soft Comput. **85**, 105862 (2019)

26. Eftimov, T., Korošec, P.: A novel statistical approach for comparing meta-heuristic stochastic optimization algorithms according to the distribution of solutions in the search space. Inf. Sci. **489**, 255–273 (2019)

27. Eftimov, T., Korošec, P.: Understanding exploration and exploitation powers of meta-heuristic stochastic optimization algorithms through statistical analysis. In: Proceedings of the Genetic and Evolutionary Computation Conference Companion, pp. 21–22 (2019)

28. Eftimov, T., Korošec, P.: Is the statistical significance between stochastic optimization algorithms' performances also significant in practice? In: Proceedings of the 2020 Genetic and Evolutionary Computation Conference Companion, pp. 19–20 (2020)

29. Eftimov, T., Korošec, P.: Data repository for the book the deep statistical comparison for meta-heuristic stochastic optimization algorithms (2021). https://doi.org/10.5281/zenodo.5767232

30. Eftimov, T., Korošec, P.: Deep statistical comparison for multi-objective stochastic optimization algorithms. Swarm Evolut. Comput. **61**, 100837 (2021)

31. Eftimov, T., Korošec, P.: Reducing bias in multi-objective optimization benchmarking. In: Proceedings of the Genetic and Evolutionary Computation Conference Companion, pp. 27–28 (2021)

32. Eftimov, T., Korošec, P.: Statistical analyses for meta-heuristic stochastic optimization algorithms. In: Proceedings of the Genetic and Evolutionary Computation Conference Companion, pp. 770–785 (2021)

33. Eftimov, T., Korosec, P., Korousic-Seljak, B.: The behavior of deep statistical comparison approach for different criteria of comparing distributions. In: IJCCI, pp. 73–82 (2017)

34. Eftimov, T., Korošec, P., Seljak, B.K.: Disadvantages of statistical comparison of stochastic optimization algorithms. In: Proceedings of the Bioinspired Optimizaiton Methods and their Applications, BIOMA, pp. 105–118 (2016)

35. Eftimov, T., Korošec, P., Seljak, B.K.: Comparing multi-objective optimization algorithms using an ensemble of quality indicators with deep statistical comparison approach. In: 2017 IEEE Symposium Series on Computational Intelligence (SSCI), pp. 1–8. IEEE (2017)

36. Eftimov, T., Korošec, P., Seljak, B.K.: Deep statistical comparison applied on quality indicators to compare multi-objective stochastic optimization algorithms. In: International Workshop on Machine Learning, Optimization, and Big Data, pp. 76–87. Springer (2017)

37. Eftimov, T., Korošec, P., Seljak, B.K.: A novel approach to statistical comparison of meta-heuristic stochastic optimization algorithms using deep statistics. Information Sciences **417**, 186–215 (2017)

38. Eftimov, T., Korošec, P., Seljak, B.K.: Data-driven preference-based deep statistical ranking for comparing multi-objective optimization algorithms. In: International Conference on Bioinspired Methods and Their Applications, pp. 138–150. Springer (2018)

39. Eftimov, T., Korošec, P., Seljak, B.K.: Deep statistical comparison of meta-heuristic stochastic optimization algorithms. In: Proceedings of the Genetic and Evolutionary Computation Conference Companion, pp. 15–16 (2018)

40. Eftimov, T., Petelin, G., Hribar, R., Popovski, G., Škvorc, U., Korošec, P.: Deep statistics: more robust performance statistics for single-objective optimization benchmarking. In: Proceedings of the 2020 Genetic and Evolutionary Computation Conference Companion, pp. 5–6 (2020)

41. Eftimov, T., Petelin, G., Korošec, P.: Dsctool: A web-service-based framework for statistical comparison of stochastic optimization algorithms. Appl. Soft Comput. **87**, 105977 (2020)

42. Engmann, S., Cousineau, D.: Comparing distributions: the two-sample anderson-darling test as an alternative to the kolmogorov-smirnoff test. J. Appl. Quant. Methods **6**(3) (2011)

43. Erdfelder, E., Faul, F., Buchner, A.: Gpower: A general power analysis program. Behav. Res. Methods Instrum. Comput. **28**(1), 1–11 (1996)

44. Fodor, I.K.: A survey of dimension reduction techniques. Cent. Appl. Sci. Comput. Lawrence Livermore Natl. Lab. **9**, 1–18 (2002)

45. Franklin, J.N.: Matrix Theory. Courier Corporation (2012)

46. García, S., Fernández, A., Luengo, J., Herrera, F.: Advanced nonparametric tests for multiple comparisons in the design of experiments in computational intelligence and data mining: experimental analysis of power. Inf. Sci. **180**(10), 2044–2064 (2010)

47. Garcia, S., Herrera, F.: An extension on"statistical comparisons of classifiers over multiple data sets" for all pairwise comparisons. J. Mach. Learn. Res. **9**(Dec), 2677–2694 (2008)

48. García, S., Molina, D., Lozano, M., Herrera, F.: A study on the use of non-parametric tests for analyzing the evolutionary algorithms' behaviour: a case study on the cec'2005 special session on real parameter optimization. J. Heuristics **15**(6), 617–644 (2009)

49. Girden, E.R.: ANOVA: repeated Measures, p. 84. Sage (1992)

50. Glover, F.: Future paths for integer programming and links to artificial intelligence. Comput. Oper. Res. **13**(5), 533–549 (1986)

51. Gordon, S.P.: Visualizing and understanding l'hopital's rule. Int. J. Math. Educ. Sci. Technol. **48**(7), 1096–1105 (2017)

52. Hansen, N., Auger, A., Finck, S., Ros, R.: Real-parameter black-box optimization benchmarking 2010: experimental setup. Doctoral dissertation, INRIA (2010)

53. Hansen, N., Auger, A., Mersmann, O., Tusar, T., Brockhoff, D.: Coco: A platform for comparing continuous optimizers in a black-box setting. arXiv preprint arXiv:1603.08785 (2016)

54. Hansen, N., Finck, S., Ros, R., Auger, A.: Real-parameter black-box optimization benchmarking 2009: noiseless functions definitions. Ph.D. thesis, INRIA (2009)

55. Heiberger, R.M., Neuwirth, E.: One-way anova. In: R through Excel, pp. 165–191. Springer (2009)

56. Henze, N.: A multivariate two-sample test based on the number of nearest neighbor type coincidences. Annal. Stat. 772–783 (1988)

57. Higham, N.J.: Computing the nearest correlation matrix-a problem from finance. IMA J. Num. Anal. **22**(3), 329–343 (2002)

58. Hochberg, Y.: A sharper bonferroni procedure for multiple tests of significance. Biometrika **75**(4), 800–802 (1988)

59. Holm, S.: A simple sequentially rejective multiple test procedure. Scand. J. Stat. 65–70 (1979)

60. Hsu, H., Lachenbruch, P.A.: Paired t test. In: Wiley Encyclopedia of Clinical Trials (2008)

61. Huband, S., Barone, L., While, L., Hingston, P.: A scalable multi-objective test problem toolkit. In: International Conference on Evolutionary Multi-criterion Optimization, pp. 280–295. Springer (2005)
62. Hutter, F., Hoos, H.H., Leyton-Brown, K.: Sequential model-based optimization for general algorithm configuration. In: International Conference on Learning and Intelligent Optimization, pp. 507–523. Springer (2011)
63. James, F.: A review of pseudorandom number generators. Comput. Phys. Commun. **60**(3), 329–344 (1990)
64. Jolliffe, I.: Principal component analysis. In: Wiley Online Library (2002)
65. Kämpf, J.H., Wetter, M., Robinson, D.: A comparison of global optimization algorithms with standard benchmark functions and real-world applications using energyplus. J. Build. Perform. Simul. **3**(2), 103–120 (2010)
66. Kirk, R.E.: Practical significance: a concept whose time has come. Educ. Psychol. Meas. **56**(5), 746–759 (1996). https://doi.org/10.1177/0013164496056005002
67. Knowles, J., Thiele, L., Zitzler, E.: A tutorial on the performance assessment of stochastic multiobjective optimizers. Tik Rep. **214**, 327–332 (2006)
68. Korošec, P., Eftimov, T.: Insights into exploration and exploitation power of optimization algorithm using dsctool. Mathematics **8**(9), 1474 (2020)
69. Korošec, P., Eftimov, T.: Multi-objective optimization benchmarking using dsctool. Mathematics **8**(5), 839 (2020)
70. Korošec, P.: Stigmergy as an approach to metaheuristic optimization. Ph.D. thesis, Jožef Stefan International Postgraduate School, Ljubljana, Slovenia, Ljubljana, Slovenia (2006). Slovenian title: Stigmergija kot pristop k metahevristični optimizaciji
71. Köster, J., Rahmann, S.: Snakemake-a scalable bioinformatics workflow engine. Bioinformatics **28**(19), 2520–2522 (2012)
72. Lam, F., Longnecker, M.: A modified wilcoxon rank sum test for paired data. Biometrika **70**(2), 510–513 (1983)
73. Lehmann, E.L., Romano, J.P., Casella, G.: Testing Statistical Hypotheses, vol. 150. Wiley, New York (1986)
74. Lilliefors, H.W.: On the kolmogorov-smirnov test for normality with mean and variance unknown. J. Am. Stat. Assoc. **62**(318), 399–402 (1967)
75. López-Ibáñez, M., Dubois-Lacoste, J., Cáceres, L.P., Birattari, M., Stützle, T.: The irace package: iterated racing for automatic algorithm configuration. Oper. Res. Perspect. **3**, 43–58 (2016)
76. Mann, H.B., Whitney, D.R.: On a test of whether one of two random variables is stochastically larger than the other. Annal. Math. Stat. 50–60 (1947)
77. Martello, S., Toth, P.: Bin-packing problem. Knapsack problems: algorithms and computer implementations, pp. 221–245 (1990)
78. Mattozzi, M.: Cocoarestclient. https://mmattozzi.github.io/cocoa-rest-client/. Accessed: 2021-10-20
79. McKight, P.E., Najab, J.: Kruskal-wallis test. In: The Corsini Encyclopedia of Psychology, pp. 1–1 (2010)
80. Metaheuristics Network.: (2006). http://www.metaheuristics.net/
81. Nemenyi, P.: Distribution-free multiple comparisons. In: Ph.D. thesis. Princeton University (1963)
82. Nobile, M.S., Tangherloni, A., Rundo, L., Spolaor, S., Besozzi, D., Mauri, G., Cazzaniga, P.: Computational intelligence for parameter estimation of biochemical systems. In: 2018 IEEE Congress on Evolutionary Computation (CEC), pp. 1–8. IEEE (2018)
83. Nocedal, J., Wright, S.J.: Numerical Optimization. Springer, New York, Berlin, Heidelberg (1999)
84. Osman, I.H., Laporte, G.: Metaheuristics: a bibliography. Annal. Oper. Res. **63**, 513–623 (1996)
85. Pearson, E.S., Agostino, R.B., Bowman, K.O.: Tests for departure from normality: comparison of powers. Biometrika **64**(2), 231–246 (1977)

86. Pettitt, A.: Testing the normality of several independent samples using the anderson-darling statistic. Appl. Stat. 156–161 (1977)
87. Pettitt, A.N.: A two-sample anderson-darling rank statistic. Biometrika **63**(1), 161–168 (1976)
88. Pošík, P., Baudiš, P.: Dimension selection in axis-parallel brent-step method for black-box optimization of separable continuous functions. In: Proceedings of the Companion Publication of the 2015 on Genetic and Evolutionary Computation Conference, pp. 1151–1158. ACM (2015)
89. Reinelt, G.: Tsplib'a traveling salesman problem library. ORSA J. Comput. **3**(4), 376–384 (1991)
90. Riquelme, N., Von Lücken, C., Baran, B.: Performance metrics in multi-objective optimization. In: Computing Conference (CLEI), 2015 Latin American, pp. 1–11. IEEE (2015)
91. Schilling, M.F.: Multivariate two-sample tests based on nearest neighbors. J. Am. Stat. Assoc. **81**(395), 799–806 (1986)
92. Schultz, B.B.: Levene's test for relative variation. Syst. Zool. **34**(4), 449–456 (1985)
93. Shapiro, S.S., Wilk, M.B.: An analysis of variance test for normality (complete samples). Biometrika **52**(3/4), 591–611 (1965)
94. Silverman, B.W.: Density Estimation for Statistics and Data Analysis. Routledge (2018)
95. Škvorc, U., Eftimov, T., Korošec, P.: Cec real-parameter optimization competitions: Progress from 2013 to 2018. In: 2019 IEEE Congress on Evolutionary Computation (CEC), pp. 3126–3133. IEEE (2019)
96. Škvorc, U., Eftimov, T., Korošec, P.: Gecco black-box optimization competitions: progress from 2009 to 2018. In: Proceedings of the Genetic and Evolutionary Computation Conference Companion, pp. 275–276 (2019)
97. Škvorc, U., Eftimov, T., Korošec, P.: Understanding the problem space in single-objective numerical optimization using exploratory landscape analysis. Appl. Soft Comput. **90**, 106138 (2020)
98. Steel, R.G., Torrie, J.H., et al.: Principles and procedures of statistics. In: Principles and Procedures of Statistics (1960)
99. Stützle, T.: Local search algorithms for combinatorial problems-analysis, algorithms and new applications. Ph.D. thesis, Darmstadt University of Technology, Department of Computer Science, Germany (1998)
100. Székely, G.J., Rizzo, M.L.: Testing for equal distributions in high dimension. InterStat **5**, 1–6 (2004)
101. Tian, Y., Peng, S., Zhang, X., Rodemann, T., Tan, K.C., Jin, Y.: A recommender system for metaheuristic algorithms for continuous optimization based on deep recurrent neural networks. IEEE Trans. Artif. Intell. **1**(1), 5–18 (2020)
102. Tušar, T., Filipič, B.: Differential evolution versus genetic algorithms in multiobjective optimization. In: International Conference on Evolutionary Multi-Criterion Optimization, pp. 257–271. Springer (2007)
103. Van Veldhuizen, D.A., Lamont, G.B.: Multiobjective evolutionary algorithm research: a history and analysis. Tech. rep, Citeseer (1998)
104. Voß, S., Martello, S., Osman, I.H., Roucairol, C.: Meta-Heuristics: advances and Trends in Local Search Paradigms for Optimization. Kluwer Academic Publishers, Dordrecht, The Netherlands (1999)
105. Wasserman, G.S.: Probability and statistics for modern engineering (1991)
106. Winer, B.J., Brown, D.R., Michels, K.M.: Statistical Principles in Experimental Design, vol. 2. McGraw-Hill, New York (1971)
107. Wolpert, D.H., Macready, W.G.: No free lunch theorems for optimization. IEEE Trans. Evolut. Comput. **1**(1), 67–82 (1997)
108. Zitzler, E., Thiele, L.: Multiobjective evolutionary algorithms: a comparative case study and the strength pareto approach. IEEE Trans. Evolut. Comput. **3**(4), 257–271 (1999)

Index

© The Editor(s) (if applicable) and The Author(s), under exclusive license to Springer   131
Nature Switzerland AG 2022
T. Eftimov and P. Korošec, *Deep Statistical Comparison for Meta-Heuristic Stochastic Optimization Algorithms*, Natural Computing Series,
https://doi.org/10.1007/978-3-030-96917-2

Printed in the United States
by Baker & Taylor Publisher Services